MINDING ◧ ANIMALS

MINDING
ANIMALS

Awareness, Emotions, and Heart

MARC BEKOFF

with a Foreword by Jane Goodall

OXFORD
UNIVERSITY PRESS

OXFORD

UNIVERSITY PRESS

Oxford New York
Auckland Bangkok Buenos Aires Cape Town Chennai
Dar es Salaam Delhi Hong Kong Istanbul Karachi Kolkata
Kuala Lumpur Madrid Melbourne Mexico City Mumbai Nairobi
São Paulo Shanghai Taipei Tokyo Toronto

First published by Oxford University Press, Inc., 2002
First issued as an Oxford University Press paperback, 2003
198 Madison Avenue, New York, New York 10016

www.oup.com

Oxford is a registered trademark of Oxford University Press

Library of Congress Cataloging-in-Publication Data
Bekoff, Marc.
Minding animals : awareness, emotions, and heart /
by Marc Bekoff; with foreword by Jane Goodall.
p. cm.
Includes bibliographical references (p.) and index.
ISBN 0-19-515077-5(cloth) ISBN 0-19-516337-0(pbk.)
1. Animal behavior. I. Title
QL751 .B366 2002
591.5—de21 2001051341

1 3 5 7 9 8 6 4 2

Printed in the United States of America
on acid-free paper

To my parents, two wonderful beings,
for their unrelenting support and
unconditional and unbridled love
as I playfully pursued my dreams
with raging optimism,
often to their bafflement

To Jethro, a trusting soul and a mindful dog
of few barks who embraces all that
is good, a being who is at peace
with himself and all the world
and truly loves all other beings

CONTENTS

FOREWORD

Jane Goodall

Marc Bekoff has titled this beautifully written, scholarly, and exciting book *Minding Animals*. He is a scientist—a cognitive ethologist—with all the right credentials: a Ph.D. from a respected university, numerous papers published in respected scientific journals, a position on the faculty of another respected university. In addition to all this, he believes that animals have rich emotional lives. And he loves them and *admits* that he loves them. He admits it in his public speaking; he admits it in this book. He truly wants us to *mind* about animals. And he provides page after fascinating page of carefully researched scientific evidence, interspersed with vivid descriptions of the encounters he has had with a wide range of animal species, all pointing to the complexity of animals' behavior, minds, and feelings. Controversial issues are discussed clearly and in depth, and the reader is allowed to make up his or her own mind.

Because Marc believes that animals, like ourselves, have rich personal lives, he is written off as "flaky" by some of his peers. Yet Marc insists that, even though it may be difficult for science to prove beyond all doubt that animals have emotions, it is equally impossible for science to prove that they do not. So animals should be given the benefit of the doubt. Marc's stand, his willingness to speculate about issues of animal consciousness and self-awareness, is courageous, for he is speaking within the citadel of academia, where, to maintain his position, he must publish in scientific journals papers that have been subjected to rigorous peer review. What is encouraging is the number of journals that *have* accepted Marc's articles on animal emotions. Slowly the attitude of more and more scientists, of more and more of the general public, is changing. The behaviorists, those who believe animals are little different from machines, preprogrammed to respond predictably to a variety of stimuli, are gradually losing ground to those who take a more commonsense attitude to what and who animals are.

In 1960 I began my own battle to gain scientific acceptance for concepts that were shocking in academia at that time: that animals had personalities, minds, and emotions. Unlike Marc, I had no scientific reputation to maintain: Louis Leakey had sent me to study chimpanzees without even a university degree. In fact he chose me because my mind had not become biased by the reductionist thinking of most ethologists at the time. Instead he chose someone who had learned about animals from a series of animal teachers from her childhood. Figaro and his successor, Pickles, were both neutered tomcats, as different from each other as chalk from cheese. Figaro loved sitting on my lap and would follow me around, while Pickles, though he would purr when you stroked him, maintained a certain aloofness. Budleigh and Rusty were dogs of very different personality. Like many children, I wanted to teach the dogs tricks. Buds learned slowly, if at all. Rusty only needed a couple of lessons to catch on. Rusty could play games like hunt-the-slipper and hide-and-seek. If I reprimanded him for something he had learned that was not permitted, he apologized, but if, in his books, he had done nothing wrong, he sulked, facing the wall. Not until I knelt beside him and apologized to him did he cheer up. Rusty was the best teacher in the field of animal behavior any child could have. Imagine the shock when I went to Cambridge University, after studying our closest living relatives in the field, and was told that I should have numbered, not named, the chimpanzees, and that it was not appropriate to talk about *nonhuman* animals having personalities, minds, or emotions. I couldn't believe this could be true. And, when film of the Gombe chimpanzees was subsequently released, I think it did a great deal to change the way people thought about the relationship between ourselves and the rest of the animal kingdom. It began to seem, as most people have always known, that there was, after all, no sharp line—as drawn by Western science and many religious denominations—between us and them, between humans and other animals.

All of this is elegantly and persuasively argued in this book, and the arguments are based on a wealth of scientific data. And although there is still a good deal of skepticism regarding animal emotions, rationality, and consciousness, ethology has, as Marc points out, moved forward: 40 years ago many scientists were not even prepared to describe an animal in terms of gender (they were all "it"). Consciousness and self-awareness is not an easy area to research, even in humans, but Marc does not take the view, held by some of his colleagues, that because it may be impossible to prove one way or the other it is not worth researching. In his quest to explore these fascinating issues Marc has teamed up with philosophers who investigate uncharted regions of the human mind and spirit, where it is equally difficult to obtain empirical data.

And he has teamed up with his dog, his close companion Jethro—just as I once teamed up with Rusty. Jethro has helped Marc ask the right questions, and together they have conducted many experiments. *Minding Animals* is made intensely readable and lively because it is illustrated by so many anecdotes—single observations of fascinating behavior. Marc shares my passionate belief that anecdotes can be extremely useful as indications of just what animals are capable of, as springboards for launching rigorous research. As he observes different animal species—and he has had firsthand experience with many—Marc is continually asking himself: what would it *be* like to be this coyote, this wolf, this dog? The question once led to an experiment with his partner Jethro. It took place as the two went for walks in the freshly fallen snow. We have all, I expect, seen responsible people, in the built-up Western world, cleaning up behind their dogs, scooping their poop into plastic bags. But Marc is gathering up Jethro's *pee*, the yellow-stained snow, and moving it elsewhere. Why? You will find out in chapter 1.

This book, because it provides deep insights into animal nature, challenges the reader to think more deeply about the way we treat animals today. Even if we only suspect that they *might* have rich emotional lives, that they *might* have self-awareness, that they *might* be capable of mental as well as physical suffering, then we cannot fail to be deeply disturbed by the way our own species inflicts so much suffering on so many animal species. There is massive exploitation of animals around the world—intensive farming, trapping for fur, hunting for "sport," testing for medical research and the pharmaceutical industry, use in the circus, advertising, and other forms of "entertainment," the pet industry, and so much more. How did the world get to be this way? The indigenous people of the world respect the animals they share their lives with. In North America they refer to them as Brothers and Sisters. They hunt for food, but they offer up a prayer of thanks to an animal they have killed. The animals lived wild and free before their death. It is very different today. And in some ways science is to blame. In medical research, for example, scientists have been only too willing to capitalize on the obvious biological similarities between chimpanzees and ourselves—similarities in the structure of DNA, in composition of blood, in responses of immune systems, and so on—and for these reasons use their living bodies to try to learn more about human diseases. Yet even though they have been used as "models" for trying to learn about human psychological disorders, most scientists have been reluctant to admit even the *possibility* of animal personality, rational thought, and emotions similar to our own. After all, as Marc points out, it must surely be easier to conduct painful invasive research on subjects that the experimenter believes feels no pain, have no emotions, no minds. Of course, we humans do torture each other, but usually after distancing ourselves psychologically; "dehumanizing" the enemy is standard practice around the world. The "enemy" is comprised of military units—so many guns, tanks, battalions, and so on. Never individual human beings, with their own names, personalities, likes and dislikes, families. The Jews, and other victims of Hitler's holocaust, were given numbers, depersonalized.

Animals are depersonalized when science labels them with numbers, when agribusiness breeds them in factory farms treating them as living machines for turning plant protein into animal protein. The traditional farmer knows his animals, and at least some individuals in a herd or flock stand out and are named. The books by James Herriott that describe his veterinary work among the hill farmers of the Yorkshire Dales are filled with descriptions of the idiosyncrasies of individual cows and pigs, goats and sheep. But the slaves in our factory farms, hormone- and antibiotic-filled prisoners being fattened up (or made lean) for the table, have no man-given names, and their personalities have no chance to express themselves. We do not wish it. The very thought that the glimmerings of rationality and personality can survive in such conditions lays on us too much guilt. Yet there are countless descriptions of animals rescued from the most horrendous imprisonment who, given a chance to be themselves or given love or space or understanding, reveal their own unique personalities. *Minding Animals* sets all this in a scientific framework. That is why it is such an important book. Too often, those who care passionately about animal suffering are written off as sentimental simply because they have no scientific background. Marc's research, along with the courageous way he shares not only his findings but also his beliefs, gives credibility to the stance of animal activists. And it will change the way that many people regard animals.

Another important section of *Minding Animals* discusses how an uncaring attitude has so often been fostered in our children. I was lucky, for my early fascination with animals, common to most children, was nurtured and guided by a very perceptive mother. When I was 18 months old, she found me in bed with a handful of earthworms. Instead of scolding me, she just said quietly, "Jane, if you keep them here they'll die. They need the earth." I gathered up the worms and toddled with them into the garden. Thus her gentle wisdom guided my early exploration of the animal world. Children quickly learn from those around them, especially those they love and those they admire. They are fascinated by animals, but just as they can learn to be kind to them, to accept them as wonderful beings, each important in its own right, so too can they learn to treat them as objects, put in the world for the benefit of humans. And they can learn to be cruel. Many children learn that the animals chosen to share their homes are to be loved, but that mice and rats and spiders should be killed along with other "pests." They discover that it's okay to kill animals for food or for their skins. Some children may learn that it is manly to shoot them for sport. Above all, children are often told that animals don't have feelings like ours, don't feel pain in the same way. This is how teachers persuade sensitive students to kill and dissect an animal in class. So our children typically come to accept the status quo. And even if they want to protest the system, how, in the face of so much opposition, can they really effect change? Marc shows what can be done, how children can, and should, be encouraged to have the courage of their convictions. He can say this because he himself dropped out of courses that required him to kill animals.

Marc is actively helping the Jane Goodall Institute develop and expand its environmental and humanitarian program for youth, as he describes in chapter 1. He is working with groups of elementary students and senior citizens, and he is extremely successful: he loves to *join* in their activities, get his hands dirty, and have fun with them. And although he has the courage of his convictions, he is neither sanctimonious nor strident. He teaches the children that, in the real world, all is not black and white, and that progress often results from a series of compromises.

We can never hope, in the foreseeable future, to eradicate all cruelty to animals, including human animals. Unfortunately we seem to have inherited certain brutal tendencies from our ancient primate heritage. But we have also inherited tendencies for compassion, altruism, and love. Once a sufficient number of people are aware of animal suffering, we can strive, realistically, to bring to an end the mass cruelty that is, implicitly or explicitly, currently condoned by society as a whole. There are few people who honestly believe that animals have no feelings. But huge numbers of people are brainwashed into accepting cruel practices because that is the way things are. They become numbed, "all pity choked by custom of fell deed." Others try to hide away because they cannot bear the suffering but lack the will to try to do anything about it. Some want to help but don't know how. The thing is that we live in a consumer-driven society, and we can help to bring about change by making ethical choices in what we buy—and, even more important, what we do not buy. There are more and more cruelty-free products on the market; we must seek them out and share our knowledge with our friends. We can insist on buying only free-range eggs. We can eliminate the cruel use of animals in circuses and other forms of entertainment by refusing to go to performances of this sort and by turning off our TV sets. We can refuse to buy dogs or cats from puppy mills or pet shops, and instead adopt unwanted animals from shelters. More and more alternatives to the use of live animals in pharmaceutical product testing are being found; we must insist on legislation that enforces their use, once they have proven safe and effective. The list of simple ways that we can, collectively, ensure change is endless. And *Minding Animals* will go a long way to opening people's minds to the fact that animals can and do suffer. Together, Marc and I have founded a new organization, Ethologists for the Ethical Treatment of Animals/Citizens for Responsible Animal Behavior Studies (www. ethologicalethics.org). It is devoted to stimulating wide-ranging discussion about the responsible and ethical conduct of behavioral research. It provides a perfect forum for exchanging ideas about some of these issues.

Read *Minding Animals* carefully. It is written for anyone and everyone who is interested in animals, for it is important that all different sectors of our society learn as much as possible about the other beings with whom we share this planet. An understanding of animals should pervade our spiritual and religious selves. If you are a student, it will help you to direct your future research, or your future lifestyle, in an ethical way. If you are someone who wants to help animals by better understanding who and what they are, this book will provide you with a feast of valuable information.

Above all, *Minding Animals* is a glorious celebration of the wonderful and diverse animal species that enrich our lives. The joyous descriptions of animals at play, revealing their spiritual nature, will inspire many to watch animals with new enthusiasm—those with whom we share our homes and backyards as well as those living in wilderness areas. And watching them, finding new questions to ask, thinking up new ways of answering them can, Marc asserts, provide not only fascinating information but also enormous pleasure, tremendous fun. He is absolutely right.

PREFACE

From the animals' point of view: Bridging the gap in the first person

Early every morning I take a nice and easy stroll with my buddy, Jethro, along Boulder Creek, near my mountain home. This is "his time," and I follow him and let him do what he wants to do. Jethro is a very large part German shepherd, part rottweiler whom I had the good fortune of meeting at the Boulder Humane Society. He is very laid-back and trusting, a passionate and well-mannered soul who is at peace with himself. Jethro is a dog of few barks, but when he speaks it behooves me and others to listen well, for his messages are drenched with deep insights into, among other matters, human nature. I let him speak freely, for I am ultimately his (and other animals') voice in matters concerning his life and I want to know what he has to say. His language is richer and deeper than mere words.

Many ideas come to me early in the morning as I listen to birds sing, the occasional coyote howl, and the water in Boulder Creek rush by. A resident family of red foxes frequently shows itself, skunks greet me with their pungent odor, mule deer casually browse just outside my kitchen window, and if I am lucky I catch a glimpse of a wandering black bear or mountain lion. I try to sense the world through the eyes, ears, and noses of these amazing nonhuman animal beings (hereafter animals).

I also commute to and from the university on my bike, and one day when I was riding home "it all came to me." Who and what go around do indeed come around. I realized that I have been writing my book since I was about four years old or perhaps even before. I always have been interested in what

Jethro, listening to me read aloud some sentences from my book, before yawning and continuing to wonder what it's all about.

(Photograph by Barbara Puskas)

animals were thinking and feeling and never doubted that they had minds and brains of their own. Thus the title of this book, *Minding Animals*. My parents have told me that I always "minded animals." I would invariably ask them what a dog was thinking, and my father recalls that on a ski trip I once asked him what a red fox was feeling when he crossed our path as we traversed a frozen lake. I use the phrase "minding animals" in two ways. First, "minding animals" refers to caring for other animal beings, respecting them for who they are, appreciating their own worldviews, and wondering what and how they are feeling and why. The second meaning refers to the fact that many animals have very active and thoughtful minds. Often I feel that Jethro "minds" me as he goes about doing what he does.

For nearly three decades I have lived in the mountains outside Boulder, Colorado. I willingly share the surrounding land with many animal friends — coyotes, mountain lions, red foxes, porcupines, raccoons, black bears, and a wide variety of birds, lizards, and insects, along with many dogs and cats. They have been my teachers and healers. They have made it clear to me that they were here first and that I am a transient on their turf. I have almost stumbled into mountain lions and have watched red foxes playing right in front of my office door. Adult bears and their young have played outside my kitchen window. I feel blessed to have had these and other experiences, and if I need to make changes in how I live to accommodate my friends, it is just fine with me.

While I can only account fully (or almost so) for my own experiences, I would be surprised if my encounters with the world-at-large were uniquely mine. Working closely with animals in the scientific arena, and also enjoying their company in this venue and at my home, I have learned much about who I am and where I fit on this amazing planet. The dynamic parameters often change by the second. The borders between "them" and "us" also are dynamic and at best blurred and permeable. There are both similarities and differences that need to be understood and appreciated. As you will see as you read on, "we" are not "them" and "they" are not "us," although there are many basic likenesses.

In addition to writing about science, animal behavior from a broad taxonomic basis, evolution, and behavioral ecology, I will trespass into other arenas. I have always been interested in matters of spirit and soul and have had the good fortune of being involved in two exciting, challenging, and ongoing interdisciplinary programs, Science and the Spiritual Quest II (www.ssq.net) sponsored by the Center for Theology and the Natural Sciences in Berkeley, California, and the program on Science, Ethics, and Religion sponsored by the organization and publisher of *Science* magazine, the American Association for the Advancement of Science (AASS). At these meetings participants speak freely about science (evolutionary biology, anthropology, psychology), spirituality, theology, religion, and God, and much progress is made in addressing how science and religion can be reconciled.

Although my training was strongly scientific, I never felt that science was the only valid approach to coming to terms with the world around me. Cut-and-dried normative science is just too confining. Assumptions that science is value-free never sat well with me, for scientists are first humans and have individual agendas about everything. I did not worship science and always thought there was room for pluralism and holism. I believe that the contributions of spiritual and religious perspectives to science are important in our coming to a fuller understanding of animal behavior, in particular the evolution of social morality. I also believe that if science and scientists are to change their ways the impetus will have to come from within the halls of science rather than from, for example, theology, although it will be painfully obvious that I believe that interdisciplinary discussions and cooperation are essential for producing change. Indeed, the new field of "neuro-theology" is growing among scholars interested in the biological bases of spirituality, meditation, and mystical and religious experiences.

Nonetheless, the philosopher Holmes Rolston III claims that "science cannot tell humans what they most need to know: the meaning of life and how to value it." I agree. Science does not allow for expressions of sentimentality or spirituality. Unchecked, science can easily produce a soul-less society and a loss of human dignity and free will. Questioning science and recognizing the limitations of scientism will make for better science by situating science in relation to other fields of inquiry in which it is more acceptable to ponder questions about spirituality, soul, life, death, grace, love, and God.

Having said all this, I want to stress that I am *not* a science-basher and that I love what I do. Questioning science is not to be antiscience nor a Luddite. I also believe that doing science should be fun and that enjoying science will make for better science. I hope that my enthusiasm for, and my love of, the study of animal behavior will be contagious, and this is among the many reasons I wrote this book. I am a "biophiliac" who loves learning about other animals and nature.

My trek to the study of animal behavior—whole animals in the field—was not a direct one. I studied neurobiology for a while, completing a master's degree on the neurochemistry of memory and part of a doctorate on visual mechanisms in cats. It was during a physiology class, when one of my professors proudly strolled into the laboratory and killed a rabbit using a rabbit punch (laughing as he did so), that my longtime interest in animal protection came to the fore. I left this program and subsequently another because I did not want to "sacrifice" cats or dogs. After moving to Washington University in St. Louis to work on my Ph.D. with the best mentor one could ever have, Michael W. Fox, I conducted research on captive animals, and I allowed mice and infant chickens to be killed by coyotes in staged encounters. I am deeply sorry and haunted by the fact that I did this sort of research, and I would never do it again. I cannot give back life to these mice and chickens, but I have anguished over their death at my hands. Another consistent thread in my research is that I have always been interested in *individuals*. Much of my research has been focused on understanding how individual differences in behavior arise during early development, and what they mean as youngsters get older and become independent. I am also interested in the evolution of behavioral variation.

There are a number of very good books that deal with different aspects of animal behavior but focus in on nonhuman primates. Given the evolutionary closeness of our primate kin, this is understandable. However, there is a vast world of other extremely interesting and talented animals who do amazing things and whose lives are filled with many mysteries that still need solving. Some of these beings, especially social carnivores such as wolves, can tell us much about the evolution of human behavior. Most primatologists pay little attention to the comparative literature on behavior, cognition, and consciousness, but I recommend that they expand their horizons, for a primatocentric view does not account for the rich diversity of animal behavior that is played out throughout the animal kingdom. Thus in *Minding Animals* I discuss the behavior of many different animals, concentrating on vertebrates. As Benjamin Beck once correctly warned his colleagues, there are dangers in being narrowly "chimpocentric."

Minding Animals: A brief road map

I have spent the last thirty-odd years studying a wide variety of animals— coyotes, wolves, dogs, Adélie penguins, archer fish, western evening gros- beaks, and Steller's jays—and tackling questions dealing with neuroethology, social development, social communication, social organization, play, an- tipredatory behavior, aggression, parental behavior, morality, function (why did some behavior pattern evolve), and the like. Studying animals, taking the time to get to know them well, is essential to gaining knowledge and feeling for how they live their lives. I want to share some of my experiences with you by telling stories and providing some of the nitty-gritty details of how my colleagues and I "make our living." In addition to personal stories, including one about my beloved mother, Beatrice Rose, I discuss the work of many of my colleagues.

In chapter 1 I tell you who I think I am and why some people think I am weird. In chapter 2 I discuss how animals are represented and misrepre- sented in the mass media and then provide a brief outline of how my col- leagues and I study animals and describe and explain their behavior so you can conduct your own studies if you so choose. In chapter 3 I present vi- gnettes of interesting behavior patterns in various species from compara- tive, evolutionary, and ecological perspectives, including predatory and antipredatory behavior, grooming and gossip, what animals know about what others know, self-medication (zoopharmacognosy), feeding patterns, dreaming, dominance, mating behavior including sexual selection, sperm wars, and mate choice. I hope that after reading these chapters you will agree that the study of behavior, ethology, is not a "soft" science and that students of behavior do not have to apologize to "hard" scientists (for ex- ample, chemists and physicists) for "merely watching animals." (Indeed, as many of my students point out, ethology *is* a "hard" science because it is so difficult to study behavior under field conditions.) And I bet we have more fun than they do anyway!

I then turn to some "big" areas in chapters 4, 5, and 6. In the fourth chapter I discuss animal minds and what is in them, animal cognition, in- telligence, and consciousness, and in the fifth I focus on animal emotions.

Throughout I argue that it is not very useful to ask if cats are smarter than dogs or chimpanzees are smarter than wolves, for each individual has to do what she or he needs to do in her or his own world. Individuals of many different species are highly talented, and broadening studies of cognition and intelligence to species other than nonhuman primates is essential if we are to learn about the evolution of variations in cognitive abilities in different species. We need to don our thinking caps, use our own big brains, and develop clever ways to study questions about what animals know about what others think and feel.

I also believe that it is not of great concern if some or all animals are not self-conscious—that they do not know *who* they are—for they may not need to be able to represent themselves to themselves in their worlds, and thus this capacity did not evolve. In many cases, recognition that "this is my body" may be necessary and sufficient for individuals to make differentiations between "self and other" and to function as "card-carrying" members of their species. Innovative new studies are needed to assess the presence or absence of self-consciousness in a broader array of species. Perhaps we have been looking for self in all the wrong places.

Concerning animal emotions, I stress that for many animals the real question of interest is not *if* they have emotions but rather *why* emotions have evolved and what functions they might serve. There are also notable individual differences in "personalities" and temperaments within species and marked variations among species. In chapter 5 I provide some background about the nature of animal emotions, present some vivid and compelling examples of animal passions, and suggest how we might go about learning more about the deep emotional lives of our animal kin. There is much exciting and very challenging work to do in this area. We have barely broken the surface from which we can delve into animals' passionate natures.

Next, in chapter 6, I move on to a detailed discussion of social play behavior and argue that social play might be a "foundation of fairness" and can provide insights into the evolution of social morality—what is permissible and what is not permissible during social interactions. It is essential to pay close attention to the nuances and the dance of play, for cooperative social play is a serious matter. During play, animals assess themselves and others, fine-tune ongoing interactions, and attempt to negotiate agreements so that play can continue. Many authors who write about the evolution of social morality fail to consider play in their discussions, and I believe this is an oversight that needs to be remedied in the future. I also argue that it may feel good to be nice and to behave fairly, that these feelings might play a role in the evolution of cooperation, and that we need to play more daily. Cooperation is not always merely a by-product of tempering aggressive and selfish tendencies (combating selfish genes) and attempts at reconciliation. Rather, cooperation and fairness can evolve on their own because they are important in the formation and maintenance of social relationships. This view, in which nature is sanitized, contrasts with those who see aggression, cheating, selfishness, and perhaps amorality as driving the evolution of sociality.

In chapters 7, 8, 9, and 10 I stray into even bigger areas and perhaps onto thinner ice. I also move into practical matters, for there is a "real world" out there that needs to be dealt with, and I believe that all scientists have broad and deep social responsibilities. In chapter 7 I discuss matters of animal protection, focusing on differences between the views of animal welfarists and animal rightists. I also discuss the various ways in which animals are used by humans for predominantly anthropocentric—humanocentric—ends. In chapter 8 I discuss how researchers can influence the very animals they study and produce misleading data. If we are to draw reliable inferences from our work, we must be sure that we are not influencing the animals to the point that we are unable to answer the very questions in which we are interested. While some of the questions that I raise bring up difficult issues about how we conduct some research and whether we should even do some studies at all, I maintain that the harder the questions, the harder (or better) the science.

In chapter 9 I continue my discussion of human intrusions into nature, concentrating on projects in which humans attempt to "redecorate" or reorder nature by moving animals from place to place in efforts to "restore" ecosystems so as to maintain or to increase biodiversity. I also discuss "science" and argue that we need a more socially responsible, compassionate, and holistic science if we are ever to make progress solving the problems that we have created all over our planet. Finally, in chapter 10, I try to tie it all together.

I conclude that love is the answer. I know that this sounds like a lot of fluff to some people and that talk of love often makes people feel very uncomfortable. Regardless, I sincerely believe that we must intimately connect with and love other animals, other humans, and all environments if we are to continue to live with grace and in harmony on this wondrous and interconnected planet. We need animals, and we need wildness and wilderness, to be healthy human beings. While I was writing this book I heard a newscast and then read a report from the Associated Press that emphasized just how interconnected we all are. Scientists from the U.S. Geological Survey discovered that dust from the Sahara Desert is blown across the Atlantic Ocean and found in the Caribbean and in the United States. This dust carries with it tiny microbes that survive the five- to seven-day journey. Bacteria, fungi, and viruses actually hitch a ride across the ocean and find their way to locations thousands of miles away. As a result, there is a heightened risk of respiratory diseases. One study found that on a dusty day there was an average of 158 bacteria and 213 viruses in a quart of air, whereas on a clear day there was an average of only 18 bacteria and 18 viruses in the same volume of air.

The notion of "intimate interrelationship" is very important to me. I long for the time when all life is woven into a seamless tapestry—a soft and warm blanket—of respect, compassion, and love. As I write below, when I study coyotes I am coyote; when I study penguins, I am penguin. I undergo the transformation in as complete a manner as I can, feeling their joys and pains and suffering. For me, this empathic connection is essential and primal and forces me to confront issues about who I am and what I do during my short time on Earth. Yes, I do get tired of aching with others' pains, being stilled and stunned and saying "I'm sorry" and begging for forgiveness when nonconsenting animals suffer, but it is my own anguish along with the bountiful joy that animals bring to me that makes me who I am. I immensely dislike being an apologist for human arrogance. If we change our ways, apologies will become matters of the past.

There are innumerable and looming "bio-realities" reflecting the damage that we have done and are doing to our one and only planet. We are a dynamic force in nature with an intensity far greater than any other previous animate force. We must do something now to reverse and to stop the destruction for which we are responsible. Determination and wisdom are needed. We need to be proactive and compassionate activists now, and not when it is "more convenient." I never cease to be perplexed by the paradox that we know how we wreak havoc on Earth and the universe and then continue doing so as if nothing has happened. I think that this repeated pattern of intrusion and destruction happens for at least two reasons. First, it works in the short term, and many people do not have long-term visions of what the future will be like for their children and their children's children. It also works because it is easy to say something like "Someone else will do it, someone else will pick up after my mess, I'm too busy." We are *all* responsible, and none of us can any longer disregard our moral and civic responsibilities. We need to learn to coexist with all other animal beings, with trees and plants, with inanimate landscapes, and with air and water. We are *all* citizens of Earth. We must never forget this as we head into the future. Time is not on our side. It really isn't!

Minding Animals has consumed me in many pleasant ways. It rarely was a burden, and I slowly became the book. I wrote it in my head, on a pad beside my bed, and even, on occasion, got off my bike and called my phone machine to leave a message with ideas, some of which obviously were a result of exercise-induced anoxia. Stick'ems were stuck everywhere, and I would later discover them in places that I could not even imagine. I do not need much more than five hours of sleep a night, and this helped. As my dear (and prolific) friend Michael Tobias once told me when I questioned him about his working hours—I receive his e-mails from all over the globe at the oddest of times—"I'll sleep when I die." Maybe I will and maybe I will not, but it is of no matter now for there is much fun to be had in doing the important and time-consuming "work" that needs to be done by all of us.

Thank you for joining me on my journey. I see "my" journey as "our" journey. I am very pleased to have had the good fortune to be able to write this book, and I hope that you will not suffer from my good luck. To paraphrase what the poet e. e. cummings once said before a lecture at Harvard University back in the early 1950s, I am glad to be here and I hope you are not sorry.

When all is said and done, and usually more is said than done, I love to imagine that all nonhuman and human beings will come to live in a soulscape bounded by, and immersed in, grace, mutual compassion, respect, and love. This is how I maintain unflagging hope. I remain hopeful and have great faith that we can make this a better world for ourselves, our children, and theirs, because we *are* a very special species, but not *better* than other species. We need to step lightly, watching each and every step that we take.

Acknowledgments

It is very difficult, and in some ways impossible, to thank all of the wonderful people who have influenced me. Indeed, some might wish to go unnoticed. Nonetheless, during the preparation of this book, and in the years during which I worked out various editions in my head and on paper, there were a number of people who truly helped me along. Sharon Adams selflessly offered much-needed support in many different arenas and has gently and brilliantly challenged many of my views. My close friend and sister-insomniac Jane Goodall is always there to provide down-to-earth discussion and wisdom, and Mary Lewis patiently listened to me as I tried to find Jane as she traveled all over the world. Mary's lightness and sense of humor really made me laugh and took the edge off some otherwise gnarly days. Thomas Berry endured long telephone calls about animals, ethics, and community, and his deep and powerful insights continually moved me into numerous unanticipated places (as they have moved many others). A number of people read an ancestral draft of the entire manuscript, and as far as I know they're still sighted and sane. Colin Allen, with whom I have done much collaborative work and cycled thousands of miles in the United States and abroad, read the whole book and was always there to pull in my reins and ask, "Do you *really* mean that" or "What in the world are you trying to say?" Dale Jamieson patiently walked with me through oftentimes tortuous philosophical literature and remains a true friend, although I'd often ask him, "Do you all really spend time pondering this obvious fact?" Benjamin Beck always grounded me and made me appreciate just how dedicated are many people who work with captive animals. I remain indebted to numerous students who truly inspired me over the years. Kirk Jensen, my editor at Oxford, provided detailed comments and very useful suggestions on earlier drafts of my book. Perhaps more than anyone else he helped me see my vision more clearly than I had before he took his red pen to prose. I also thank Kim Congleton, Kate Pruss, and Joellyn Ausanka at Oxford for help with this book. India Cooper was an awesome copy editor. Jonathon Lazear, Christi Cardenas, Tanya Cromley, and Julie Mayo provided more support than one could ever expect from "their agent." Others who provided very helpful comments or guided me to information of which I was unaware include Joel Berger, Lee Dugatkin, Marc Hauser, Mike Huffman, Leslie Irvine, Gregory Peterson, Mary Thurston, and Steve Wise. Kathleen Dudzinski, Dave Mech, and Barbara Puskas kindly provided photographs. Tom Mangelsen graciously provided the photograph for the cover. My research has been funded by a variety of federal agencies and private groups, and I am grateful for their support.

And, once again, there is Jethro, who would listen to me read sentences aloud, yawn, scratch an ear, roll his eyes, and go back to sleep, but always be there when I needed him.

MINDING ◧ ANIMALS

one

CHASING COYOTES AND MOVING "YELLOW SNOW"

Two happy foxes make my day

On the morning I began working on chapter 3, soon after spending four hours searching for stick'ems that were stuck to other stick'ems, looking for papers and books in my cluttered office (anyone who has seen it knows this is a gross understatement), and relentlessly tapping the keys on my keyboard, I rode to the university on my bike. The path I chose goes up Four Mile Canyon, over a steep dirt road (Poorman Road for those who know Boulder) surrounded by ponderosa pine trees where I have had the pleasure of meeting many deer, coyotes, squirrels, birds, and friendly dogs, and then down Sunshine Canyon, where I can descend at upwards of fifty miles per hour and enjoy the wind on my face. Luckily, I never have seen a mountain lion or a black bear on Poorman Road, but I carry pepper spray with me because lions and bears are around and I am sure that on more than one occasion one of them has watched me struggle on the steep incline. A former student of mine, an All-American cross-country skier, was once chased up a tree by a female lion who was protecting two cubs. Lucky for Linda she was in such good physical shape.

As I was climbing on the dirt road, I looked ahead and saw a small red fox running down the road on my left. He stopped, urinated, and then continued on his merry way. His tail was high and wagging, and his gait was light and frisky. Then, immediately on my right, I saw another red fox, whose tail was going around like a propeller and who was emitting almost inaudible high-pitched whines. The foxes came together on the run and greeted one another effusively. They licked one another's muzzles, their tails wagging so rapidly they could have become airborne, their whines a melodious crescendo, and then they took off over the side of the road. One of my students said, when I told him about this encounter, "You're one lucky dude." He is correct. Seeing these happy foxes made me feel great and healed all the mental strife I had experienced looking for this paper or that paper or a book that I had long given away. Just the previous month I had seen a fox bury another fox near my house. Animals can indeed be healers, and how fortunate I was to have such a natural remedy for a hectic morning. I truly felt blessed.

When Shirley met Jenny: Long-lost friends

Elephants have strong feelings. They experience joy, grief, and depression, and they mourn the loss of their friends. Elephants live in matriarchal societies in which strong social bonds among individuals endure for decades. They also have great memory. Shirley and Jenny, two female elephants who were unintentionally reunited after living apart for twenty-two years, showed that they truly had missed one another when they were separated. At different times, each was brought to the Elephant Sanctuary in Hohenwald, Tennessee, founded and run by Carol Buckley, so that they could live out their lives in peace, absent the abuse they had suffered in the entertainment industry. Upon their initial meeting, when Shirley was introduced to Jenny, there was an urgency in Jenny's behavior. She wanted to get into the same stall with Shirley. Loud roars emanated from deep in each elephant's heart as if they were old friends. Rather than being cautious and uncertain about one another, they touched one another through the bars separating them and remained in close contact. Their keepers were intrigued by how outgoing each was. A search of records showed that Shirley and Jenny had lived together twenty-two years before in the same circus when Jenny was a calf and Shirley was in her twenties. They still remembered one another, as individuals, when they were inadvertently reunited.

Maternal care and mother love in killer whales

Killer whales, or orcas as they are called, are born ready to swim. They are very curious as infants and will do just about whatever their mothers and siblings will allow them to do. Naomi Rose, who works at the Humane Society of the United States and who has studied the behavior of whales for many years, once watched a large group of orcas along the coast of Vancouver

A mother bottlenose dolphin and her calf swim in an echelon. In these dolphins and many other animals, mothers and their young form very close social bonds.

(Photograph by Kathleen Dudzinski)

Island in Canada. One day Naomi saw a mother surface without her calf and discovered that the young orca had swum behind her boat and his head was very close to the engine. The calf was playing and having a good time exploring his surroundings. In a few minutes, the calf's mother came close to the boat as her calf played. She tolerated his curiosity and playful spirit, while remaining close if her child should need her. She watched him closely but let him enjoy himself. Being overprotective is not necessarily a good thing, because youngsters have to make some mistakes as they grow up and learn "the rules of the road." Perhaps orca mothers also know this. They clearly are closely bonded to their children, but part of maternal care and mother love is shown by allowing children to make their own mistakes and being there to offer support when they need it.

Grieving for Gemer

Dolphins form long-term social bonds. They play, travel, and forage together. Denise Herzing, a professor of biology at Florida Atlantic University, has been studying Atlantic spotted dolphins for almost two decades. Three of the first dolphins she met were Rosemole, Little Gash, and Mugsy. In 1990, both Rosemole and Mugsy were pregnant, and Denise expected to return to her field site the following spring to see their infants. But when she arrived back in the field in spring 1991, only Rosemole had a child, Rosebud. Mugsy swam on the periphery of the mother-calf group, child-less, looking despondent. She swam in line with the other mothers, leaving a space below her as if she had a phantom calf in tow. She was listless and showed little interest in her friends or in mating opportunities. Denise was touched by the extent of Mugsy's grief for her lost calf.

Denise also observed another dolphin mother, Gemini, grieve over the loss of her calf, Gemer. Gemini had been attacked by a shark, and when Denise saw her she was emaciated, despondent, and disoriented. Gemer was never seen again, likely a victim of the attack. Gemini recuperated and subsequently gave birth to two other calves. Grief in dolphins and many other animals is experienced and processed, and the individuals then move on to do what they need to do to survive, including providing a new chance at life for their other youngsters.

Chimpanzee revenge: Franz and Larry

Ron Schusterman, a marine biologist at the University of California, Santa Cruz, has a story that shows just how crafty an animal can be. A young male chimpanzee, Franz, who years ago lived at the Yerkes Laboratory of Primate Biology, was known to be a feces thrower. Ron's friend Larry was one of Franz's favorite targets. One day, Larry noticed that Franz's cage had been cleaned up and teased Franz, "You can't get me—na na na na na." Franz stared at Larry while he was being taunted. When Larry finished, Franz regurgitated partially digested food he had been fed a few minutes earlier and threw it at Larry. Larry was splattered, and Franz ran around in a victory dance.

A hiker's incredulity

"Hey, what in the world are you doing?" asked an incredulous hiker on the bike path near my mountain home.

"Why, I'm just scooping yellow snow and moving it around to see how my dog, Jethro, responds to his own and other dogs' urine," I answered, shrugged my shoulders, and continued scooping and moving.

"Oh, I see," said the hiker, and then he looked at me as if I were a bit odd and walked away, no doubt relieved that he would not have to interact with this weirdo anymore. I don't think that he really did see at all. I wondered why he was so surprised. Wouldn't everyone understand that it is just part of a normal day to move urine-packed snow from place to place?

No, not many people would understand. Only a few people are lucky enough to be able to study nonhuman animals and to do the things that I and many of my colleagues do. Not many people ask, "What is it like to be a particular animal?" and then design studies that take them deeply into the worlds of other animals living in beautiful places that are often very inaccessible and harsh. It is a never-ending journey. Just when I think that I know all there is to know about some aspect of behavior, I discover that there are many surprises in store. My sister, Margie, once asked me the deceptively simple question, "How do you know when you know an animal?" I thought about her question for a long time and realized that I never really "know" an animal in the sense of knowing all there is to know. However, with each investigation and "interview" that my colleagues and I conduct, we do indeed unravel more and more of the mystery and awe of what it is like to be the animal we are studying. It is similar to unwrapping the skin of an onion, but in this case there is no final layer and then nothingness. I have found that there *always* is something more to learn.

What I love about what I do is that there are all sorts of ways to go about studying animal behavior. It is very challenging to enter the worlds of other animals, and some ingenuity often is needed. One of my former graduate students, the noted behavioral ecologist and conservation biologist Joel Berger, once disguised himself as a female moose by donning a "moose suit" so that he could get close enough to moose who were not habituated to humans to study their responses to different odors, in order to learn about how they reacted to odors of past and current predators. (The results of Joel's clever and important research project will be discussed in chapter 3.)

Into the hearts of animals

For more than thirty years I have had the good fortune to study animal behavior and to welcome many animals into my life. What an exciting and adventurous journey it has been. I work in beautiful environs, ask questions, and gather data to answer them, and the answers generate many more questions; there never is a dull moment. It is indeed a privilege to study animals.

After I left a graduate program in neurobiology and behavior at a major medical school because I did not want to sacrifice dogs or cats as part of my education, I discovered what I really wanted to do. I wanted to study social behavior in animals and not have to dispose of ("sacrifice") the animals when I was done, individuals with whom I bonded. My parents were a bit disappointed that I had decided that my graduate education would not involve getting two degrees, a Ph.D. and then an M.D., but rather only a Ph.D. They would have to settle for listening to their friends tell them that it was nice I was a doctor, but not, you know, a "real doctor." But as the years rolled on, my parents and their friends saw how much I love what I do and also realized that by studying animals I could learn much about humans. For them and others, the human connection was important.

So what's a Jewish boy from New York doing moving "yellow snow" on a cold Colorado day to learn why dogs urinate where and when they do, or watching Adélie penguins in Antarctica raise their young and try to avoid being eaten by south polar skuas, leopard seals, and killer whales? Why have I braved sixty degrees below zero and driving snowstorms to observe coyotes in Wyoming as they went about courting, mating, raising young, and finding and defending food and territory? Why have I sat on the roof of my house filming western evening grosbeaks to learn about how they detect and avoid predators, and what could possibly have driven me to watch dogs, coyotes, and wolves playing with one another and wondering if they trust one another and if they are moral beings? And surely there are better things to do than pondering whether nonhuman animals are "persons" or thinking about animal spirituality and wondering how studying other animals—learning about them, understanding them, and appreciating and loving them for *who* they are (and not *what* they can do for us)—can tell me about human nature?

It really is legitimate to ask what someone with my upbringing is doing writing this book. I often ask myself the same question, for I was raised in a home where most of the talk centered on sports, school activities, or whether there was a new Three Stooges episode on TV. I did not grow up in the country, nor was I raised with animals. I lived with some goldfish when I was young, and I always wondered how they liked being in their fish tank. But I felt a lot of compassion and love in my home, and I know that this is related to the fact that I have always loved animals. Indeed, the title of this book, *Minding Animals*, stems from a conversation that I had with my parents in the late 1980s. They told me that I always wanted to know what animals were thinking and feeling. I don't remember it, but they said that when I was about four years old I yelled at a man for hitting his dog, and the man chased after my father! My father recalled that when we were on a cross-country ski trip in Pennsylvania when I was five, a red fox ran in front of us as we broke trail, and I was absolutely taken with the beauty of the animal and immediately asked questions about where the fox lived and whether he was happy. I still ask such questions and plan to forever.

One thing is very clear. If I had had to predict what I would be doing now, it wouldn't have been what I am in fact doing! The major message is that in many ways it is not worth worrying about where you will be in fifty years; indeed, it is probably not worth worrying much about where you will be in fifty seconds. Surely, making plans is essential, but if I hadn't been willing to rethink mine, I might have spent far too much time and energy trying to do something for which I was not suited. I love what I do, being immersed in nature and having the privilege of studying animal behavior. I consider myself very fortunate to have had the opportunity to have so much fun while making a living. I have been able to follow my heart, to bring animals into my heart, and to travel into their hearts.

My research has taken me in many different directions. Most important, it has led me deep into the minds, hearts, spirits, and souls of many other animals. It has also led me deeply into my own mind, heart, spirit, and soul. Animals have been my teachers and healers. Animals are a way of knowing.

So I am sitting here scratching my head and rubbing Jethro's stomach, writing this book and trying to figure out why. While I cannot answer with much certainty the question of how I came to do the things that I do, I do know that my heartfelt passion and my curiosity about the world, always encouraged by my parents, is at the root of many of the seemingly weird activities in which I participate as an ethologist who studies the lives of other animals. I also love teaching and working closely with students as they unravel the mysteries of the awesome and magnificent world we are all so lucky to live in.

Socially responsible science, compassion, and heart

> My prayer is that we "center down," for the sake of all the relations, for all of us. To be perfectly honest—and there can be nothing less—my prayer is that we *get down*, that we get down and dirty. I pray that we lose ourselves while lovemaking with dirt, with the rocks and streams, the salmon who swim there, the coyotes and 'coons, the water bugs and snakes—with the fertile ground of wherever we may be.
> —Laura Sewall, *Sight and Sensibility*

"Oh, you're acting just like an animal." Many people suffer this indignity but casually shrug it off with "Thanks for the compliment." I do too, fairly often. I do not see anything negative about being called "an animal." I hope you will agree with me when you have finished this book.

I love studying other animals. I love them for who they are, beings with lives that I want to understand and appreciate. I identify with the animals I study. But I also do rigorous science, and I believe that "hard" science, socially responsible science, compassion, heart, and love can be blended into a productive recipe for learning more about the fascinating lives of other animals and the world within which each of us lives. Many scientists pay lip-service to this, often dismissing those who want to imbue science with compassion, heart, and love as "flakes" or "new-agers." In my opinion, their narrow views of science are extremely outdated. Anton Moser, who studies biotechnology at the University of Graz in Austria, is of the opinion that the "soft" and "hard" sciences can be unified to produce "deep science" in which aesthetic and sentient experiences are merged with deduction, induction, and conventional scientific practices. I agree and will follow up on this line of reasoning when I argue that studies of behavior are important in tapping into human spirituality. While we live in a "scientific world," science cannot do it alone. We need help from other disciplines, including those whose concerns are with theology and spirituality. We need to seek bravely and boldly a compassionate new world. We need a new paradigm that is replete with compassion and love, one in which compassion spills over into ethical responsibility to other animals and to the planet as whole.

I also realize, as do many of my colleagues, that science is not value-free. All individuals have a view of the world, and it is important not only to recognize this but also to develop ways to integrate individual subjectivity and values with scientific "objectivity" so as to produce the most reliable results possible.

In studies of behavior it is much more difficult not to identify or empathize with another animal than it is for a geologist not to identify or empathize with a rock or a botanist not to identify or empathize with a tree. Thus the study of animal behavior is especially challenging because it is clear that viewing animals in the cold confines of the halls of traditional science with objectivity that cannot truly be achieved can be as misleading as viewing them and getting all wrapped up in their emotional lives. While there are many similarities between many nonhuman animals and humans, there also are many differences among all animals. The similarities and the differences must be given careful attention. It is simplistic and misleading to assume that "they" are merely "us" in disguise, and I hope this message is clear as you read about the fascinating lives of our animal kin. One of my major goals in *Minding Animals* is to go beyond nonhuman primates and to expose you to the wonderful and mysterious worlds of many different animals.

Minding animals and deep ethology

The phrase "minding animals" means two things. First, "minding animals" refers to caring for other animal beings, respecting them for who they are, appreciating their own worldviews, and wondering what and how they are feeling and why.

Minding animals led me to develop the term "deep ethology" (in a discussion with my friend Michael Tobias) to convey some of the same general ideas that underlie the "deep ecology" movement, in which it is stressed that people need to recognize that they not only are an integral part of nature but also have unique responsibilities to nature. As a deep ethologist, following the tradition of ecopsychology, I, as the "seer," try to become the "seen." I become coyote, I become penguin. I try to step into animals' sensory and locomotor worlds to discover what it might be like to be a given individual, how they sense their surroundings, and how they behave and move about in certain situations.

As big-brained and omnipresent mammals, we have enormous social responsibilities: to conduct our studies in the most ethical manner, to share our information with nonscientists, and also to seek and to use their input in trying to determine which questions are the most pressing in our pursuits of knowledge. We also need to seek out the advice of those who take nontraditional views—for example, those of indigenous people who have lived for eons in close association with wild animals and other nature. There are many different ways of knowing, and all are important to consider. In the past, many Western scientists have marched into other countries, studied their exotic wildlife, and departed without consulting with the local people or giving much attention to local problems. This practice has changed greatly as Westerners have incorporated local people into their work and have applied their findings to help solve local problems. Information on many aspects of behavior—movement patterns, social organization, reproductive habits—is useful to people worldwide who are studying conservation biology and wildlife management. There is a very practical side to the "heady" study of animal behavior.

It is also important to note that there is ample evidence that compassion begets compassion and that cruelty begets cruelty. There is a close relationship between cruelty to animals and cruelty to humans. The Humane Society of the United States has a program called First Strike devoted to looking at this very thing. Developing an understanding of, and a deep appreciation for, animals is one way to begin the journey of making this a more compassionate world. In addition, one practical advantage of appreciating animals is that changes in heart might lead to less resistance to preserving critical habitat for many endangered animals whose future existence on this earth is seriously imperiled. Habitat loss is considered by most conservation biologists to be the biggest threat to animal and plant biodiversity. *Caring might indeed spill over into sharing.*

The more we come to understand other animals, the more we will appreciate them as the amazing beings they are. Who would have thought that rats might dream about the mazes in which they ran the previous day, that bonobos can tell others where they have gone by leaving signs for their friends to follow, that chimpanzees know what others know, that elephants and whales can communicate with one another over distances of a hundred or more kilometers, or that a parrot can learn to differentiate among different objects based on their shape, color, or texture?

Studying animal behavior is more than merely stamp collecting

Animals are amazing beings who display very sophisticated patterns of behavior, and this is why I am so excited to share some aspects of their wondrous lives with you. It is important to teach through understanding. I will also show you that animal behavior is not a "soft" science and that it need not be a side-show to, or a casualty of, the "hard" sciences such as physics or chemistry. Students of behavior use many sophisticated analytical techniques. As Colin Allen and I noted in *Species of Mind*, in these days of big-budget, high-technology science the very idea that observations with the naked eye or a pair of field glasses could launch and maintain a branch of science, ethology, seems almost laughable. Yet many of today's ethologists, if they carry more than field glasses and notebooks, still for the most part rely on relatively low-tech items such as audio recorders, stopwatches, hand counters, telemetry equipment, and cameras. The mating preferences of bower birds in Australia have been studied using colored poker chips!

Because of this low-tech approach, the study of animal behavior can mistakenly appear to be a simple task compared to other scientific endeavors. How hard, after all, could it be just to sit and watch animals "do their thing"? Those who make a living studying animal behavior know that studies of behavior can be rather difficult. That is what is so appealing and challenging to field workers who devise clever studies to learn about animals who differ from us in how they sense the world and move around in it. The examples of animal behavior that I discuss later on will illustrate what I am talking about.

Minding the blurred borders between "animals" and "humans"

It is very clear that learning about other animal beings—how they spend their time, who they interact with, where they do what they do and how they do it, their intellectual and cognitive abilities (cognitive ethology), and their emotional lives—is essential for gaining a full appreciation of human spirituality and what it is to be human. Just what is it that is unique about humans? Researchers have now discovered that tool use, language use, and self-consciousness, culture, art, and rationality no longer can reliably be used to draw species boundaries that separate humans from other animals. But reflecting on one's own mortality seems to be uniquely human. (I sometimes wonder if, and worry that, sadism is a uniquely human characteristic.) Claims that only humans use tools or language, are artists, have culture, or reason are no longer are defensible given the enormous growth in our knowledge of our animal kin. Primatologists have identified about forty different behavior patterns that show cultural variation in chimpanzees (tool use, grooming, patterns of courtship). Female killer whales are known to spend years showing their youngsters how to hunt elephant seals according to local custom. Researchers have compiled a list of almost twenty behavior patterns in cetaceans that are influenced by local tradition and show cultural variation. Frans de Waal, a primatologist at Emory University, tells a story of how enamored some art critics were of a painting only to change their minds when they discovered that a chimpanzee was the artist.

David Haberman, who teaches at Indiana University and conducts research in India, told me a fascinating story of a behavioral tradition that spread among the rhesus monkeys living in the temple town of Vrindaban, about a hundred miles south of Delhi. The large population of rhesus monkeys is protected, and individuals fiercely compete for food. In the early 1990s a monkey learned that if he stole eyeglasses from a human he could get food—bananas and roasted chickpeas—for returning them. This behavior spread rapidly; soon numerous monkeys were stealing glasses to obtain food, and people who lived with the monkeys learned to remove their glasses lest they lose them and have to barter for them. People who were unaware of this rhesus tradition quickly learned that they too could not wear their glasses without being robbed of them.

Although the borders between many nonhuman animals and humans are blurred and permeable, there are numerous differences as well as similarities. We need to celebrate both the diversity and the likenesses, for it is both similarities and differences that excite and challenge me and my colleagues in our research. It is also important to emphasize that little is to be gained from comparing, for example, the cognitive or emotional "level" of a chimpanzee to that of a human child, for each individual does what he or she needs to do to adapt to the demands in his or her own world. Some investigators claim that an adult chimpanzee functions at about the level of a two-and-a-half-year-old human child, but surely a young child could never survive in the chimpanzee's world. We must seek to understand each and every individual in his or her own world and be extremely cautious of thinking of differences in terms of their being "good" or conferring more "value" on an individual's life.

Animals as "persons": Beatrice and Jethro

Here is a personal story about my mother that raises many questions I will consider throughout this book. "Personhood" is a topic that has been increasingly pursued by philosophers, legal scholars, and a handful of anthropologists, psychologists, and biologists. There are practical as well as theoretical—ivory tower—issues at stake, for how we view animals, their moral and legal standing (are they objects or property or beings?), is often translated into how we treat them. Discussing the status of animals—that is, whether nonhuman animals can be considered to be persons—compels us to consider what makes us human. The study of animal cognition and emotions is central to questions about personhood.

Once, when I was visiting my parents, my father asked, "Marc, can you please wheel Mom into the kitchen and get her ready for dinner?" I answered, "Sure, Dad" and began the short trek. But the journey went well beyond the confines of my parents' home. It remains a difficult and multidimensional pilgrimage for which there are not any road maps or dress rehearsals. I watch myself watching Mom. The role reversal is riveting; I am now my keeper's keeper. I keep wondering: where (and who) is the person I called "Mom"?

My mother, Beatrice Rose, whom I love dearly, has suffered major losses of locomotor, cognitive, and physiological functions. She does not know who I am and likely has lost some self-awareness and body-awareness. She has become, as the legal scholar Rebecca Dresser calls such humans, a "missing person." In a nutshell, my mother has lost her autonomy. She has little self-determination. Nevertheless, there is no doubt others would still think of her as a "person" whose spirit and soul are very much alive and who is entitled to certain moral and legal standing. And in my view they should.

Generally, the following criteria are used to designate a being as a "person": being conscious of one's surroundings, being able to reason, experiencing various emotions, having a sense of self, adjusting to changing situations, and performing various cognitive and intellectual tasks. While many humans fulfill most if not all of these criteria, there are humans who do not, notably young infants and seriously mentally challenged adults. But they are also rightfully considered to be persons.

Now, what about my companion dog, Jethro? He is active, can feed and groom himself, and is very emotional. Jethro is as autonomous as a dog can be. Yet many people would not feel comfortable calling Jethro a "person." This irreverence would be a prime example of just what is wrong with academic musings!

Why are there different attitudes toward my mother and Jethro? Why are some people, especially in Western cultures, hesitant to call chimpanzees, gorillas, dolphins, elephants, wolves, and dogs, for example, "persons," even when they meet the criteria for personhood, more so than some humans? Perhaps it is fear. Many people fear that elevating animal beings to persons would mean that the notion of personhood is tarnished, that it means less for humans. Some also fear that animals will then have the same legal and moral standing as humans and that they will be equals. I do not consider my mother and Jethro equals.

While some may believe this whole exercise is shamefully crass, there are some very important issues at stake. Loving Jethro (and other animals) as much as I do does *not* mean I love my mother (or other humans) less. Granting Jethro and other animals personhood and attendant moral and legal standing does *not* lessen nor take away humans' moral and legal standing. In my view, such fears are not warranted.

I believe that little is to be gained by claiming that granting personhood to some animals would be a misguided or blasphemous move. Surely, Jethro goes through life differently from most human (and other dog) beings, but this does not mean he does not have any life at all. People vary greatly. There are countless different personalities, but the term *person* is broad enough to encompass and to celebrate this marvelous diversity.

Calling a nonhuman a person does not degrade the notion of personhood. However, this move would mean that animals would come to be treated with the respect and compassion that is due them, that their interests in not suffering would be given equal consideration with those of humans. I hope to convince you that nothing is lost by calling some animals "persons" and allowing all human beings to be called "persons" as well.

Animal emotions

If there is one aspect of animal behavior that attracts many people and also informs them as to how they believe humans should interact with animals, it is animals' unmistakable and easy-to-recognize emotions, their raw and unfiltered expressions of a wide variety of feelings. People always seem to want to know more about how animals experience their emotions, intense joy and deep grief, embarrassment, resentment, and love.

Numerous stories in my book *The Smile of a Dolphin: Remarkable Accounts of Animal Emotions*, written by world-famous scientists, show clearly that many animals have deep emotional lives. My colleagues' stories also show that the scientists themselves had emotional feelings for the animals they studied. The study of animal emotions raises many challenging questions. As examples and to introduce the subject, I will consider grief, joy, love, and sibling rivalry (aggression and anger) here; I discuss animal emotions in greater detail in chapter 5.

Animal emotions are usually easily recognizable. Many animals display profound grief at the loss or absence of a close friend or loved one. Jane Goodall, the world-renowned primatologist and conservationist who has studied chimpanzees for more than forty years, observed Flint, a young chimpanzee, withdraw from his group, stop feeding, and die of a broken heart after his mother, Flo, died. Flint remained for several hours where Flo lay, then struggled on a little further, curled up, and never moved again. Sea lion mothers watching their babies being eaten by killer whales wail pitifully, in anguish of their loss. Dolphins have been seen struggling to save a dead infant and mourn afterwards. And elephants have been observed standing guard over a stillborn baby for days with their heads and ears hung down, quiet, and moving slowly. Joyce Poole, who has studied wild elephants for almost two decades, believes that grief and depression in orphan elephants are real phenomena. Orphan elephants who saw their mothers being killed often wake up screaming.

On the flip side, animals also experience immense joy when they play, greet friends, groom one another, or are freed from confinement, and perhaps while they watch others having fun. Joy is contagious. Animals tell us they are happy by their behavior: they are relaxed, walk loosely as if their arms and legs are attached to their bodies by rubber bands, smile, and go with the flow. They also speak in their own tongues, purring, barking, or squealing in contentment. Dolphins chuckle when they are happy. Greeting ceremonies in African wild dogs involve cacophonies of squealing, propeller-like tail wagging, and bounding gaits. When coyotes or wolves reunite they gallop toward one another whining and smiling, their tails wagging wildly. Upon meeting they lick one another's muzzles, roll over, and flail their legs. They are jubilant. When elephants reunite there is a raucous celebration. They flap their ears, spin about, and emit a "greeting rumble." They are very happy to see one another.

Joy also abounds in play. Animals get so immersed it can be said that they *are* the joy and the play. They show their delight by their acrobatic movements, gleeful vocalizations, and smiles. There is also a feeling of incredible freedom in the flow of play. Violet-green swallows soar, chase one another, and wrestle in the grass outside my office window. I saw a young elk in Rocky Mountain National Park run across a snow field, jump in the air and twist his body while in flight, stop, catch his breath, and do it again and again. Buffaloes have been seen playfully running onto and sliding across ice, excitedly bellowing *gwaaa* as they did so.

There is accumulating evidence that animals and humans share many of the same neurochemicals that underlie emotional feelings in humans. Dopamine, serotonin, and norepinephrine, neurotransmitters associated with enjoyment and pleasure, regulate play, and opioids are related to feeling relaxed or "socially comfortable," a condition important for facilitating play.

The more we study animal emotions and the more open we are to their existence, the more we learn about the fascinating emotional lives of other animals. Surely it would be narrow-minded to think that humans are the only animals who have evolved deep emotional feelings. Indeed, evolutionary biology warns us that this cannot be the case.

A lion, a fox, and a funeral

I recently saw a female fox bury a male fox, perhaps her mate. One morning when I went out to hike with Jethro, I looked down my road and saw a female red fox trying to cover the carcass of a red fox who had been killed by a mountain lion two days earlier. I was fascinated, for she was deliberately orienting her body so that when she kicked debris with her hind legs it would cover the carcass. There has been a family of foxes near my house for almost a decade, and I assumed that she was related to, or at least a close friend of, the deceased. She kicked dirt, stopped, looked at the carcass, and intentionally kicked again. I observed this "ritual" for about twenty seconds. A few hours later I went to see the carcass and found that it was now totally buried.

No one to whom I have spoken, naturalists or professional biologists, has ever seen a red fox bury another red fox. However, a retired realtor who read my essay about it in the local paper called me and told me he had seen a red fox bury another fox in upstate New York in the 1970s. I later learned that a similar observation had been made for wolves. I do not know if the female fox was intentionally trying to bury her friend, but there is no reason to assume she was not. Perhaps she was grieving, as domestic dogs, elephants, chimpanzees, and other animals do, and I was observing a fox funeral. In 1947 a naturalist on the East Coast saw a male fox lick his mate as she lay dead. He also vigorously protected her. Perhaps the male was showing respect for a dead friend.

I mention this story because a number of people expressed skepticism when I told them that I had observed this behavior in a fox. When I asked them if they would have a different attitude if the fox had been a chimpanzee, some said yes and that they might consider more seriously that the survivor was grieving. One person told me that she would even speculate about the religious experience that might have been felt by the survivor if she had been a primate, but surely not by a fox.

I was lucky to have this series of encounters, for nature does not hold court at our convenience. Much happens in the complex lives of our animal

kin to which we are not privy, but when we are fortunate enough to see animals at work, how splendid it is. Long live natural history!

A dog and a bunny

Jethro has always been low-key, gentle, and well mannered. He never
chases animals and loves to hang out and watch the world around him. He
has been a perfect field assistant for me as I studied various birds, western
evening grosbeaks and Steller's jays, living near my house.

One day, while I was sitting inside, I heard Jethro come to the front
door. Instead of whining as he usually did when he wanted to come in, he
just sat there. I looked at him and noticed a small furry object in his mouth.
My first reaction was "Oh no, he killed a bird." However, when I opened
the door, Jethro proceeded to drop at my feet a very young bunny, drenched
in his saliva, who was still moving. I could not see any injuries, only a small
bundle of fur who needed warmth, food, and love. Jethro looked up at me,
wide-eyed, as if he wanted me to praise him for being good-natured. I did.
He was so proud of his compassionate self. I guessed that the bunny's
mother had disappeared. Most likely she fell prey to a coyote, red fox, or
the occasional mountain lion around my house.

When I picked the bunny up, Jethro got very concerned. He tried to snatch
her from my hands, whined, and followed me around as I gathered a box, a
blanket, and some water and food. I gently placed the bunny in the box,
named her Bunny, and wrapped her in the blanket. After a while I put some
mashed-up carrots, celery, and lettuce near her, and she tried to eat. I also
made sure that she knew where the water was. All the while, Jethro was standing behind me, panting, dripping saliva on my shoulder, and watching my
every move. I thought he would go for Bunny or the food, but rather he stood
there, fascinated by this little ball of fur slowly moving about in her new home.

When I had to leave the box I called Jethro, but he simply would not
leave. He usually came to me immediately, especially when I offered him a
bone, but he steadfastly remained near the box for hours on end. Finally I
had to drag Jethro out to give him his nightly walk. When we returned he
bee-lined for the box, and that is where he slept through the night. I tried
to get Jethro to go to his usual sleeping spot, and he refused. "No way," he
told me, "I'm staying here." I trusted Jethro not to harm Bunny, and he did
not during the two weeks during which I nursed her back to health so that
I could release her near my house. Jethro had adopted Bunny; he was her
friend. He would make sure that no one harmed her.

Finally the day came when I introduced Bunny to the outdoors. Jethro
and I walked to the east side of my house, and I released her from her box

and watched her slowly make her way into a woodpile. She was cautious. Her senses were overwhelmed by the new stimuli—sights, sounds, and odors—to which she was now exposed. Bunny remained in the woodpile for about an hour until she boldly stepped out to begin life as a full-fledged rabbit. Jethro remained where he had lain down and watched the whole scenario. He never took his eyes off Bunny and never tried to approach her or to snatch her.

Bunny hung around for a few months. Every time I let Jethro out of the house, he immediately ran to the spot where she was released. When he arrived there he would cock his head and move it from side to side, looking for Bunny. This lasted for about six months! When I would utter, "Bunny," in a high-pitched voice, Jethro would whine and go look for her. Bunny was his friend, and he was hoping to see her once again.

I am not sure what happened to Bunny. Other bunnies and adult rabbits have come and gone, and Jethro looks at each of them, perhaps wondering if they are Bunny. He tries to get as close as he can. He never chases them.

Jethro is a truly compassionate soul. Two summers ago, nine years after he met Bunny and treated her with delicate compassion, he came running up to me with a wet animal in his mouth. Hmm, I wondered, another bunny? I asked him to drop it and he did. This time it was a young bird who had flown into a window. It was stunned and just needed to gain its senses. I held it in my hands for a few minutes. Jethro, in true fashion, watched our every move. When I thought it was ready to fly, I placed the bird on the railing of my porch. Jethro approached it, sniffed it, stepped back, and watched it fly away.

Jethro has saved two animals from death. He could easily have gulped each down with little effort. But you don't do that to friends, do you?

Sadness and sulking by coyotes

My students and I studied the social behavior of coyotes for seven years around Blacktail Butte, in the Grand Teton National Park, south of Jackson, Wyoming. A female, whom we called "Mom," began leaving her family for short forays after giving birth to a litter in each of the previous three years. She would disappear for a few hours and then return to the pack as if nothing had happened.

I wondered if her family missed her when she wandered about. It seemed that they did. When Mom left for forays that lasted for longer and longer periods of time, often a day or two, some pack members would look at her curiously before she left. They would cock their heads to the side, squint, and furrow their brows as if they were asking, "Where are you going now?" Some of her children would even follow her for a while. When Mom was gone, the other pack members were unusually quiet, often looking in the direction in which she had last disappeared. When Mom returned, they would greet her effusively by whining loudly, licking her muzzle, wagging their tails like windmills, and rolling over in front of her in glee. "Mom's back!" Their sadness instantaneously turned into elation and joy. Her children and mate had missed her when she was gone.

One day Mom left the pack and never again returned. She had disappeared. The pack waited impatiently for days and days. Some coyotes paced nervously about as if they were expectant parents, whereas others went off on short trips only to return empty-handed. They traveled in the direction she had gone, sniffed in places she might have visited, and howled as if calling her home. For more than a week some spark seemed to be gone. Her family missed her. I think the coyotes would have cried if they could.

It was clear to all of us that coyotes, like many other animals, have deep and complicated feelings. They were sad, some clearly grieving. Their behavior told it all. They walked around with their tails hanging down, heads low—moping—despondent over the loss of their beloved mother.

Animals in love

Animal love means different things to different people. Some people even believe animals do not truly feel love for one another. But it is unlikely that love appeared in humans with no evolutionary precursors, no animal lovers.

The biologist Bernd Heinrich, an expert on ravens, believes these birds fall in love. In *Mind of the Raven,* Heinrich wrote: "Since ravens have long-term mates, I suspect that they fall in love like us, simply because some internal reward is required to maintain a long-term pair bond." Raven parents have to cooperate to capture prey for their young. They remain near one another during the day and also sleep next to one another and make soft vocalizations. They play with and preen one another and share food. During courtship feeding they quietly hold each other's bills. In his book *Here I Am—Where Are You? The Behavior of the Greylag Goose,* Konrad Lorenz, who won a Nobel Prize for this work on animal behavior, observed that "the greylag goose's peculiar process of 'falling in love' in many ways resembles its human counterpart." After bonding, males and females are strongly devoted to one another.

Even in monogamous species in which the same male and female breed from year to year, courtship is prolonged and vows need to be continually renewed. In coyotes and wolves, males and females who mated previously will act as if they never have done so. While courting they sniff one another repeatedly, play, follow one another, and eventually form an exclusive unit. They also rebuff interlopers. When reuniting they greet effusively, whining and licking one another's muzzles. If another male attempts to mate with his consort, he will drive the intruder off and defend her. Likewise, the female will reject males with whom she has no interest in mating. In raccoon dogs of South America, males emit a mating cry called the "yearning call" that resembles the distress call of youngsters. Male golden jackals utter an "entreaty call" during courtship. Some ethologists consider these calls to be signs of falling in love.

In light of Heinrich's and Lorenz's studies, it is interesting to note that more than 90 percent of bird species are monogamous. In some, males and females form very close long-lasting bonds and enjoy high reproductive success. Weakly bonded geese do not produce as many young as pairs who are strongly bonded. In seagulls, males and females who mate from year to year have higher reproductive success. In strongly monogamous Bewick swans, mated pairs who remain together also produce more young than pairs that divorce.

The strength of the bond that is formed between mates is evidence that a close, enduring loving relationship has been established. Watch rejected individuals when former mates or current beaus choose another animal with whom to consort. I have seen male coyotes, after being rebuffed by a possible mate, slink away dejectedly, head and tail hanging low. Their behavior is totally different from instances when another individual has rejected them.

Romantic love can also be inferred from the profound grief that individuals display when mates disappear or die. Male greylag geese usually do not mate after losing a mate. They cannot recover from their bereavement.

Animal love may not really be any more baffling than human love. Many different notions have passed for human love, yet we do not deny its existence. And how wrong we can be about inferring love or its absence in humans. About half of all marriages in the United States end in divorce. Some people claim they love their former partners after parting. Some claim they never loved them at all. Human love is indeed confusing. Perhaps, after all is said and done, animal love is less confusing or mysterious because animals do not filter their emotions—their love is out there for all to feel if they choose to feel it.

Sibling rivalry and siblicide

Brothers and sisters can be the best and worst of friends. Brotherly and sisterly love can be absent among those individuals in which one would expect to see it bloom. Siblings will protect one another and cooperate against a common enemy. They will also evict one another from the safety of a nest, steal food from one another even in times of plenty, or kill one another without blinking an eye. Parents, who can be very nurturing, often will merely watch food theft or mortal combat, called "siblicide," and not interfere despite the eerie cacophony of protestations by their own kids who are tossed from a nest, starve to death, or are actively killed.

Doug Mock, a biologist at the University of Oklahoma, has conducted extensive studies of sibling rivalry. Many invertebrates and vertebrates engage in intense sibling rivalry. Much research has been done on birds because they are easy to study. In many birds, chicks hatch asynchronously. Older individuals are larger than younger brood mates, and birth order effects and competitive advantages arise. In European bee-eaters, it can take nine days for an entire brood to hatch.

Not all sib rivalry involves fiery battles. In American pelicans, sib fighting is common but rarely fatal. Losers usually starve to death. Red-winged blackbirds and accipiter hawks engage in scramble competition for food in which the speed of consumption results in increased survival for individuals who get it. Successful birds take the best position in the nest or elevate their bills faster and higher. In American robins, persistent beggars usually position themselves better than their sibs to get food from parents. In pigs, the mom's anterior teats produce more milk than other nipples. Immediately after birth, large piglets take and defend the anterior teats. Small piglets suckle longer on less productive teats and are more likely to be crushed by their mothers.

In addition to monopolizing food, sibs may fight to the death. This occurs in black eagles, cattle egrets, great blue herons, and Mexican blue-footed boobies. In blue-throated bee-eaters, hooked mandibles are used by nestlings to kill sibs. It is murder most fowl indeed!

Usually oldest and largest sibs start fights. In kittiwakes, senior sibs have been observed to initiate more than 98 percent of all fights. In one study of black eagles, the older chick relentlessly pursued his younger sibling and pecked him more than 1,500 times during thirty-eight attacks. The senior chick gained fifty grams during this period, whereas the younger chick lost eighteen grams.

There are few good examples of siblicide in mammals because much rivalry occurs in dens and cannot be seen. In spotted hyenas, incisors and canines are fully erupted at birth. Cubs are born about an hour apart, and the older cub commonly attacks its younger sib immediately after it is born. Same-sex aggression is more common and heated than aggression between sexes. Thus same-sex litters are uncommon three weeks after birth. In Galapagos fur seals, newborns may be killed by one- or two-year-old siblings when food is scarce.

In all cases, it is hypothesized that the survivor's reproductive fitness is increased by reducing sib competition. Information on reproductive success is difficult to collect, especially in the field. However, in western gulls high-ranking individuals in a brood show higher survival than low-ranking individuals. In jackdaws, the last-born chick is the smallest and suffers the highest mortality, and most are emaciated at death.

There are few general explanations of sibling rivalry that apply to the diverse species in which this odious and enigmatic behavior occurs. Why do parents overproduce in the first place? As morbid as it is, an understanding of fatal siblicide in nonhumans would likely help us understand why similar behavior rarely occurs in humans, yet other forms of sibling rivalry and birth order effects are common.

What a dog's nose knows: Making sense of scents

Now, what about moving "yellow snow"? I confess, I did it! I wanted to learn about the olfactory world of dogs, and moving urine-soaked snow seemed like a good way to do so. Not surprisingly, no one had ever previously done anything like this for free-running animals. But it is obvious to just about everyone that dogs spend lots of time with their well-endowed nostrils stubbornly vacuuming the ground or pinned blissfully to the hind end of another dog. They have about twenty-five times the area of nasal olfactory epithelium (which carries receptor cells) and have many thousands more cells in the large olfactory region of their brain (mean area of 7,000 mm^2) than humans (500 mm^2). Dogs can differentiate dilutions of one part per billion, distinguish T-shirts worn by identical twins, and follow odor trails; they are ten thousand times more sensitive than humans to certain odors.

When dogs wiggle their noses and inhale (suction) and exhale (snort), they concentrate odors, pool them into mixtures, and expel some. Like their wild relatives (wolves, coyotes), dogs gather much information from the symphony of odors left behind. Urine provides critical information about who was around, their reproductive condition, and perhaps their mood. Dogs expel millions of gallons of urine wherever they please (more than 1.5 million gallons, along with about twenty-five tons of feces, per year in New York City alone) and use it well.

Odors are powerful stimulants. It is said that Sigmund Freud used soup smells to stimulate clients to recall past traumas. Although Jethro (aka Hoover) enjoys visiting his veterinarian, he will show fear if he goes into an examination room where the previous canine client was afraid. Fear is conveyed via a pungent odor released by the previous dog's anal glands.

Now, what about sniffing other dogs' urine? I studied Jethro's sniffing and urination patterns along the bicycle path near my mountain home. To learn about the role of urine in eliciting sniffing and urinating, I moved urine-saturated snow ("yellow snow") from place to place during five winters to compare Jethro's responses to his own and other dogs' urine. Immediately after Jethro or other known males or females urinated on snow, I scooped up a small clump of the yellow snow in gloves and moved it to different locations.

Moving yellow snow was a useful and novel method for discovering that Jethro spent less time sniffing his own urine than that of other males or females. Other researchers have also noted that male dogs (and coyotes and wolves) spend more time sniffing the urine from other males compared to their own urine. Dogs also usually spend more time sniffing urine from females in heat compared to urine from males or reproductively inactive females.

The differences in Jethro's response to the displaced urine from other males or from females are worth noting, especially when considering "scent-marking" behavior. Scent-marking is differentiated from "merely urinating" by a number of criteria that include sniffing before urinating followed by directing the stream of urine at urine that is already known to be present or at another target. When Jethro arrived at displaced urine, he infrequently urinated over or sniffed and then immediately urinated over (scent-marked) his own urine, but he sniffed and then immediately scent-marked the displaced urine significantly more when it was from other males than when it was from females. While domestic dogs are usually not very territorial (despite myths to the contrary), their wild relatives are, and they show similar patterns of scent-marking behavior in territorial defense.

I hope this brief foray into the olfactory world of dogs removes some of the mystery of dog sniffing and gives some idea of what the dog's nose tells the dog's brain. You can easily repeat this simple experiment (and risk being called weird). The hidden tales of yellow snow reveal a lot about the artistry of how dogs make sense of scents.

Teach the children well: The dog in a lifeboat

> Teach your children what we have taught our children, that the earth is our mother. Whatever befalls the earth befalls the sons and daughters of the earth.
>
> —Chief Seattle, "The Earth Is Our Mother"

I have spent much time working with children, helping them learn about the natural world and asking them questions about the nature of their interactions with other animals. In addition to being a lot of fun and a rich learning experience for me, it is pretty easy to teach children about animals, for they are intuitive curious naturalists who sympathize and empathize with animals. They are sponges for knowledge, absorbing, retaining, and using new information at astounding rates. Children are also full of common sense that they freely share with others. We all know this about children, but often we forget when we are helping to develop their roles as future ambassadors with other animals, nature, and ourselves. Surely it is easier to make lifelong changes in individuals' attitudes toward nonhumans and other humans if codes of conduct are taught and discussed as early in life as possible, before irreversible attitudes have been established. The link between cruelty to animals and cruelty to humans also needs to be kept in mind.

I have been fortunate to teach and have mutually beneficial discussions with some fourth graders at Foothills Elementary School in Boulder. We considered such topics as animal behavior, ecology, conservation biology, and the nature of human-animal interactions. I was astounded by the level of discussion. The class centered on the guiding principles of Jane Goodall's worldwide Roots & Shoots program, whose basic tenets are that every individual is important and every individual makes a difference. The program is activity-oriented, and members partake in projects that have three components: care and concern for animals, human communities, and the places in which we all live together.

All the students had actively been engaged in projects that fulfilled all three components. They had participated in, or suggested for future involvement, such activities as recycling, being responsible for companion animals, reducing driving, developing rehabilitation centers for animals, helping injured animals, getting companion animals from humane shelters, boycotting pet stores, tagging animals so if they get lost people would know who they are, visiting senior citizen centers and homeless shelters, punishing litterbugs, and punishing people who harmed animals. We discussed how easy it is to do things that make a difference and also develop a compassionate and respectful attitude toward animals, people, and environments. One student noted that by walking and cleaning up after the companion dog who lived with his elderly neighbor, he performed activities that satisfied all three components.

Some students had already developed very sophisticated attitudes about human-animal interactions. One thought experiment in which we engaged is called "the dog in the lifeboat." Basically, there are three humans and one dog in a lifeboat, and one of the four has to be thrown overboard because the boat cannot hold all of them. Generally, when this situation is discussed, most people reluctantly agree that, all other things being equal, the dog has to go. One can also introduce variations on the theme. For example, perhaps two of

Two projects that I have coordinated as a leader of Roots & Shoots activities involve complet-
ing and drawing a picture for the following phrases: "I have a dream that ____" and "I am
thankful for ____ ." This picture was drawn at an event in Chicago. All responses show clear
concern and respect for animals, people, and the environment. Some examples:

> My dream is that animals get enough to eat.   My dream is that I don't want my
> sister to get hurt.   My dream is do not cut down trees and save the animals'
> homes.   My dream is that butterflies will be free.   My dream is that my grand-
> mother is not sad that her two brothers died.   My dream is that everyone has
> shelter.   My dream is to meet Nelson Mandela.   I am thankful for my fruit
> bowl and my house.   I am thankful for mother earth.   I am thankful for forests
> and animals.   I am thankful for my family.   I am thankful for bird houses and
> love.   I am thankful for giving.

This last thanks was accompanied by a drawing of the kindergarten student giving a homeless
person a sandwich.

the humans are healthy youngsters and one is an elderly person who is blind,
deaf, paralyzed, without any family or friends, and likely to die within a
week. The dog is a healthy puppy. The students admitted that this was a
very difficult situation and that maybe, just maybe, the elderly human might
be sacrificed because he had already lived a full life, wouldn't be missed, and
had little future. Indeed, this is very sophisticated thinking—that perhaps
the elderly person had less to lose than either of the other humans or the
dog. Let me stress that all students agreed that this line of thinking was not
meant to devalue the elderly human. In the end, the students, like most

other people, reluctantly concluded that regardless of the humans' ages or other characteristics, the dog has to go. But this decision was not an easy one to make, and what was wonderful was the concern that the students showed about all of the individuals who were involved in the dilemma.

What really amazed and pleased me was that before we ever got to discussing alternatives, all students wanted to work it out so that no one had to be thrown overboard. Why did any individual have to be thrown over, they asked. Let's not do it. When I said that the thought experiment required that at least one individual had to be tossed, they said this was not acceptable! I sat there smiling and thinking: now, these are the kinds of people in whose hands I would feel comfortable placing my future. Some ideas about how all individuals could be saved included having the dog swim alongside the boat, having them all switch off swimming, throwing overboard everything unnecessary to reduce weight and bulk (including their shoes), and cutting the boat in two and making two rafts. All students thought that even if the dog had to go, she would have a better chance of living because more could be done by the humans to save the dog than vice versa. Very sophisticated reasoning indeed. I have discussed this example many times, and never before had a group unanimously decided that everyone must be saved.

I also was thrilled by the commitment of the teachers I met. They were dedicated souls, and we should all be grateful that such precious and priceless beings are responsible for educating future adults on whom we will all be dependent.

Swimming with dolphins

A few years ago I was standing in line at a grocery store when I heard a young girl tell her friend that she had just gone swimming with dolphins when on holiday in Hawaii. She had had a great time, but there was a slight pause when her friend asked her about what the dolphins might have felt about all of this. Did they enjoy her touching them or riding on their backs? Did they really like being bothered? I smiled and walked away thinking: now, isn't that what we want, young kids asking questions about ethics even if they do not know what they are doing?

This encounter made me think deeply about how animals and humans interact and about whether we should be doing some of the things that we do, often with little thought. It also was among the reasons I decided to write a children's book on animal ethics, *Strolling with Our Kin*, and edit *The Smile of a Dolphin: Remarkable Accounts of Animal Emotions*. Maybe we adults also need to do a bit more soul-searching about our never-ending intrusions into the lives of other animals. So, for example, we might ask, should humans swim with dolphins? Even if it is not in the dolphin's best interests, is it all right if it helps you or a family member? A friend? A stranger? Should we ever intrude on other animals if our interactions do not have some positive effect on them? These nagging questions rapidly lead us into difficult terrain, the negotiation of which requires open minds and especially open hearts.

In *Strolling with Our Kin* I presented some questions that encourage people to confront head-on some very difficult issues. These included asking whether we should interfere in animals' lives when we have spoiled their habitats or when they are sick, provide food when there isn't enough to go around, stop predators in their tracks, or translocate individuals from one place to another, including zoos and aquariums. Should our interests trump theirs? Concerning dolphins, is it permissible to play music to dolphins as long as they can move away? Is it permissible to swim with dolphins if they and the human swimmers enjoy it? Is it permissible to swim with dolphins if the dolphins and humans both benefit? If only the humans benefit? If only the dolphins benefit? Is it allowable to use motorboats and to water-ski where they live? Should we interfere with serious aggressive encounters between two dolphins?

The question of when humans *should* intrude is a difficult one. However, just because we *can* do something doesn't mean that we *should* or *have* to do it. Furthermore, just because some intrusions may be *relatively* less benign than others, this sort of claim places us on a very slippery slope and can lead to thoroughly selfish anthropocentric claims. Even in situations when we have good intentions, good intentions are not always enough.

Putting animals to sleep

Our interactions with other animals often force us to tap into our own spirituality. People who have been denied contact with companion animals, who often are treated as family members, do not seem to thrive as well as people who have not had contact with them. Alan Beck, Marshall Meyers, and Clint Sanders have discovered that companion animals can be social catalysts, providing their humans with opportunities to meet people, and also allow them to be alone without feeling lonely.

It is when we make the decision to be responsible for the last breath of another individual that our "humanness" and spirituality are clearly shown. This story is a lesson in spirituality, ritual, and love. For thirteen years Anne and I shared our home with Inuk, a huge white malamute. Over and over he would say to us, "Come on, it's time for a hike, or dinner, or a belly rub." We were constantly on call for him. As Inuk got older it became clear that our lives together would soon be over. The uninhibited and exuberant wagging of his huge tail that cooled us in the summer, occasionally knocked glasses off the table, and told us how happy he was would soon stop. What should we do—let him live in misery or help him die peacefully with dignity? It was our call, and a hard one at that.

Because companion animals are so dependent on us, we are responsible for making difficult decisions about when to end their lives, to "put them to sleep." I have been faced with this situation many times and have anguished, trying to do "what is right" for my buddies. Should I let them live a bit longer, or has the time really come to say good-bye? When Inuk got old and could hardly walk, eat, or hold down food, the time had come to put him out of his misery. We stopped feeding him the horrible medicine that made him sick so he could live a few extra weeks. Instead, we fed him ice cream and cookies, and he began to thrive. But this was only a brief flash in the pan— Inuk was dying right in front of our eyes, and we knew it. Truth be told, even when eating ice cream, he seemed miserable and told us this in many ways.

Deciding when to end an animal's life is a real-life moral drama. There are no dress rehearsals, and doing it once does not make doing it again any easier. Inuk knew we would do what was best for him, and we really came to feel that often he would look at us and say, "It's okay, please take me out of my misery and lessen your burden. Let me have a dignified ending to what was a great life. Neither of us feels better letting me go on like this." Inuk clearly let us know how he felt in this situation.

The rituals of putting Inuk and other companions out of their misery— recognizing that their lives on Earth were nearing an end, thanking them for their companionship, trust, devotion, and love, telling them how much I valued and loved them, and coming to terms with the responsibilities that I took on by sharing my life with them—were lessons in spirituality and love that made me come to terms with just how mighty and omnipresent we humans are.

Redecorating nature

> The earth is, to a certain extent, our mother. She is so kind, because whatever we do, she tolerates it. But now, the time has come when our power to destroy is so extreme that Mother Earth is compelled to tell us to be careful. The population explosion and many other indicators make that clear, don't they? Nature has its own natural limitations.
> —His Holiness the Dalai Lama, *The Path to Tranquillity*

Humans are omnipresent. Give us an inch and we take a foot and more. There are no places on Earth that are not influenced by human activities. As such we are very much a part of nature, with numerous responsibilities that cannot be pushed aside because it's convenient or because there always will be someone else to clean up the messes we leave.

Because humans have incredible power to dominate animals, our animal kin depend on our goodwill and mercy. Animals depend on humans to have their best interests in mind. *We can choose to be intrusive, abusive, or compassionate.* We do not have to do something just because someone else wants us to do it. We do not have to do something just because we can do it. Each of us is responsible for our choices. Some ideals we should base them on include (1) putting respect, compassion, and admiration for other animals first and foremost; (2) taking seriously the animals' points of view; (3) erring on the animals' side when uncertain about their feeling pain or suffering; (4) recognizing that almost all of the methods that are used to study animals, even in the field, are intrusions on their lives—much research is fundamentally exploitative; (5) recognizing how misguided are speciesistic views concerning vague notions such as intelligence and cognitive or mental complexity for informing assessments of well-being; (6) focusing on the importance of individuals; (7) appreciating individual variation and the diversity of the lives of different individuals in the worlds within which they live; (8) appealing to what some call questionable practices that have no place in the conduct of science, such as the use of common sense and empathy; and (9) using broadly based rules of fidelity and non-intervention as guiding principles. We need to reconcile common sense with "science sense."

The numbers speak for themselves

In addition to global issues concerning biodiversity, conservation, and humans' attempts to redecorate nature, there are concerns that center on individual animals rather than on entire ecosystems, populations, or species. Because there are so many people on this planet, the demand for animal products and for dealing with human medical needs and food requirements is rising astronomically.

The numbers tell the grim story. Gail Eisnitz, in her book *Slaughterhouse: The Shocking Story of Greed, Neglect, and Inhumane Treatment Inside the U.S. Meat Industry,* reports that in the United States alone, at least ninety-three million pigs, thirty-seven million cattle, two million calves, six million horses, goats, and sheep, and nearly ten *billion* chickens and turkeys are slaughtered for food each year.

Numerous animals are also used in experimental research. In 1996, according to a survey by the U.S. Department of Agriculture, the number of animals used in experimentation totaled about 1.3 million individuals, including 52,000 nonhuman primates, 82,000 dogs, 26,000 cats, 246,000 hamsters, and 339,000 rabbits. This staggering number does *not* include rats, mice, and birds, about 90 percent of the animals used in experimental research, animals who are not protected from being used in experimentation. Currently, there is legislation pending to amend the federal Laboratory Animal Welfare Act to protect these animals. It is estimated that more than seventy million animals are used annually and that one animal dies every three seconds in American laboratories alone. In the United Kingdom in 1998, according to a report in the Sunday *Independent,* more than 6.5 million mice, 2.4 million rats, and 1,000 dogs were killed because they were not needed for research after being bred for such purposes. Many spent their entire lives waiting to be used before being deemed "useless" and killed.

According to the recent report *The State of the Animals 2001*, in the United States animal use is not as well documented, nor are the data on animal use as reliable, as in other countries. There appears to be a downward trend in animal use in the United States, but not as large as that seen in countries such as Switzerland, Germany, and Great Britain. There also is inconsistency in the ways in which research protocols are evaluated by Institutional Animal Care and Use Committees (IACUCs) in different universities. Some projects that are approved in one university are not permitted in others.

Government workers also kill numerous animals. For example, some people who work for the Bureau of Land Management and the U.S. Fish and Wildlife Service "recreationally shoot" and kill prairie dogs in order to control populations of these beautiful and family-oriented rodents. Wildlife Services (formerly called Animal Damage Control), a branch of the Department of Agriculture, is responsible for cruelly and indiscriminately killing (using leg-hold traps, snares, explosives, and poisons that cause much pain and suffering) hundreds of thousands of animals—varmints, as they call them—including coyotes (over 85,000 in 1999), foxes (6,200), mountain lions (359), and wolves (173) in the name of control and management. In 1999, Wildlife Services killed over 96,000 predators. However, only about 1 percent of livestock losses are due to predators; 99 percent are due to disease, exposure to bad weather, illness, starvation, dehydration, and deaths at birth. Animal Damage Control has also been responsible for negatively influencing populations of at least eleven endangered species. Nontarget animals including domestic dogs also are killed.

We also need to be deeply concerned about the well-being of our companion animals (aka "pets"). In 1994, almost 60 percent of fifty-three million households in the United States had companion animals, and more than half had more than one animal. Dogs, cats, and/or birds lived in 71 percent of households in Belgium, 63 percent in France, 61 percent in Italy, and 70 percent in Ireland.

Michael Tobias, in his book *Voices from the Underground: For the Love of Animals,* reports that according to two surveys taken in 1994, there were about 235 million companion animals in the United States, including 60 million cats, 57 million dogs, 12.3 million rabbits, guinea pigs, hamsters, gerbils, and hedgehogs, 12 million fish tanks (no estimate of numbers of fish), 8 million birds, 7.3 million reptiles, and 7 million ferrets. About $17 billion per year is spent on pet supplies. In a 1998 poll reported in *Time* magazine, 83 percent of pet owners said they would likely risk their lives for their pets.

While it is wonderful to think that the sheer number of animals indicates that people truly care about their companions, this is not so. Far too many animals breed, and there are numerous unwanted individuals. Many are ignored or abused when they become burdensome to their human companions. Many are also tortured "for fun." A number of organizations, including local humane societies, have programs directly concerned with the well-being and fate of companion animals.

Patient and compassionate activism: Indifference is deadly

> I knew that if I continued to debate politics and science—and stayed in the mind instead of the heart and the spirit—it would always be one side versus the other. We all understand love, however; we all understand respect, we all understand dignity, we all understand compassion."
> —Julia Butterfly Hill, *The Legacy of Luna*

It is an understatement to claim that animals need our help. I believe that "getting my hands dirty," getting out there and showing people the ghastly things we do to far too many animals and to Earth, is the best way to make long-lasting changes in their hearts and heads. Showing, not telling, is the most effective way to practice activism. One of my favorite calls to action is "Indifference is deadly—do something, anything, to make this a better planet." My own activism centers on getting people to think and having them tell me why they think, feel, and act the ways they do.

A wonderfully moving heartfelt discussion of the trials and tribulations of activism, and also of its innumerable fruits, is "tree sitter" Julia Butterfly Hill's autobiography, *The Legacy of Luna.* Parts of this book moved me to tears. Julia, an incredibly courageous woman, sat alone in a tall California coast redwood tree she named Luna for two years to save Luna's life. When she first climbed Luna, Julia had no intention of remaining there for such a long period of time. Julia has been an inspiration for people worldwide as an example of nonviolent activism in defense of forests, and she has made a huge difference in how people view trees and in how trees, as beings in the world, are treated. She has clearly shown people that activism works and is the youngest person to be inducted into the Ecology Hall of Fame.

As an unwavering dreamer and optimist, I often feel victimized by hope. Nonetheless, it is my passionate dream that changes in attitude and heart will ultimately bring forth harmony in the relationships between animals and humans, for nonhuman animals will forever be competing with humans, their dominant, big-brained, mammalian kin. Without a doubt, the animals are likely to lose most of these encounters as humans continue to try to redecorate (manage, control) nature for their own selfish ends.

Activism for animals has helped me tap into my own spirituality, for there are numerous costs to activism—harassment, intimidation, humiliation, and frustration—that often become personal. I have felt the effects of attempts by government officials to silence my asking questions about the reintroduction of Canadian lynx into Colorado as well as my questioning why dogs had to be killed in physiology courses so that medical students could learn about life. Such personal assaults made me dig deeply into my heart in my efforts to understand and to explain to others why I was doing what I was doing, whether it was organizing protests to save animals or partaking in candlelight vigils and prayer services for animals who had been killed. Compassionate people who push the envelope can easily engender the wrath and fear of small minds. I was once called a "flake" by some of my colleagues for my position on animal rights. I was flattered and wondered why they were taking the time to engage a flake! Surely they have better things to do with their valuable time.

Every individual counts, and that every individual makes a difference. As Margaret Mead, the renowned anthropologist, noted: "Never doubt that a small group of thoughtful, committed citizens can change the world. Indeed, it is the only thing that ever has."

The comparative study of animal behavior

In my own research on social behavior and behavioral ecology, I stress evolutionary, ecological, and developmental (ontogenetic) perspectives. I follow the lead of such classical ethologists as the Nobel Prize winners Konrad Lorenz and Niko Tinbergen. I also try to understand individual differences within species and variations among species. Individual differences in behavior are exciting to study because variation provides information that highlights just how different individuals, even closely related individuals, can be. Variation is not noise to be dispensed with.

My approach is called the "comparative approach" to the study of behavior. I have done much interdisciplinary work, working with geneticists, anatomists, and philosophers. While we know much about the lives of other animals, there remain enormous gaps in our knowledge that need to be filled before we can make any hard-and-fast general claims about the evolution and development of most behavior patterns. Caution surely is the best road to take when offering generalizations, especially about complex behavior patterns, animal thinking, and animal emotions.

I also work at many different levels of analysis. While much of my research is done at the "micro" level (for example, analyzing frame-by-frame films of animals at play or animals looking out for potential predators), I am an interdisciplinary holist at heart. I prefer to tackle "big" questions. I do not shy away from conducting detailed statistical analyses, but never do the animals I am studying get thrown aside as numbers, unnamed variables in an equation, or points on a graph. It is important that the "protective membrane of statistics," as Mary Lou Randour calls it, not shield us from the worlds of other animals, their joys and pains, their wisdom, their uniqueness.

While a number of people have contributed to the foundations of the study of animal behavior, Charles Darwin's ideas were the most important contributions during the third quarter of the nineteenth century. He appears to be the first person to have applied the comparative evolutionary method to the study of behavior, in his attempt to answer questions concerning the origin of emotional expression. Darwin used six methods to study emotional expression; some of them did not work well, and others seem naive nowadays: (1) observations of infants; (2) observations of the insane, who, when compared to normal adults, were less able to hide their emotions; (3) judgments of facial expressions created by electrical stimulation of facial muscles; (4) analyses of paintings and sculptures; (5) cross-cultural comparisons of expressions and gestures, especially of people distant from Europeans; and (6) observations of animal expressions, especially those of domestic dogs.

Winning the "prize": Lorenz, Tinbergen, and von Frisch

In 1973, Konrad Lorenz, Niko Tinbergen, and Karl von Frisch shared the Nobel Prize for Physiology or Medicine. I, like many of my colleagues, was delighted. Each had made seminal contributions to the study of behavior, and finally ethology, the field of science concerned with the biological bases of behavior, was squarely on the map. Many people were aghast that scientists who spent their time "just watching animals" could out-compete biomedical researchers for what is commonly called the "prize" within academia.

Lorenz followed in Darwin's footsteps. Trained as a physician, comparative anatomist, psychologist, and philosopher, Lorenz was the first biologist to appreciate fully the importance of behavioral characters—what animals do—for taxonomic endeavors in which species are classified into their phylogenetic (evolutionary) groups. He stressed that behavior patterns could be viewed as structures, similar to stomachs, hearts, and kidneys, or as phenotypes, expressions of genes, on which natural selection acted. Thus behavior was something that an animal "did" as well as something that an animal "had."

Although behavior can be highly variable, it is, in some instances conservative enough to be used to study evolutionary relationships among different species. For example, Lorenz found that the ways in which ducks and geese perform different types of social displays—how they move their head, wings, and tail—can be used to determine which species are closely or distantly related to one another. Similarly, some colleagues and I discovered by studying the development of social play behavior and aggression that a mysterious animal called the New England canid was a combination of wolf and coyote genes, but that it was more coyote than wolf. This carnivore appeared in New England in the late 1920s, and people were not certain of its origins. It turns out that the New England canid was a cross between coyotes from Minnesota who had migrated across eastern Canada and, along the way, mated with wolves and a variety of domestic dogs. Our findings agreed with studies that analyzed measurements of skulls and serological (blood) characteristics of these hybrids, showing that behavior could be used as reliably as morphological structures and physiological parameters to study animal taxonomy. John Gittleman, at the University of Virginia, has discovered that the chemical signals used by members of the cat family (Felidae) can also be used as phylogenetic characters.

Lorenz also demonstrated how animals perceive only parts of their environments, called "key stimuli." Behavioral responses are due to a process called perceptual filtering, as the result of which some stimuli are perceived and others are not.

Lorenz is best known for his classical studies in imprinting, in which he discovered that geese would follow just about any individual with whom they were reared. Thus Lorenz found himself the father figure for the geese and other waterfowl with whom he shared his home. His work on imprinting emphasized how innate (inborn) and acquired (learned) components of behavior are integrated. Lorenz also pondered the origins of human behavior and wrote much on aggression and war. In his classic book *On Aggression*, he claimed that much human aggression was inborn.

According to his colleague Niko Tinbergen, the central point in Lorenz's work was his clear recognition that behavior is part and parcel of the adaptive equipment of animals. Behavior could be observed, described, measured, and studied quantitatively, although Lorenz did not collect much empirical information. Rather, he freely used anecdote and anthropomorphism, stressed that it was important to empathize with nonhumans, and believed that animals had the capacity to love, be jealous, experience envy, and be angry. Lorenz focused mainly on description rather than experimentation. Darwin too thought that scientists often underestimated the mental abilities of other animals, and he also freely used anecdotes to make his case. Darwin attributed cognitive states to many animals on the basis of observation of particular cases rather than controlled experiments.

Niko Tinbergen is often called "the curious naturalist." Tinbergen worked with Lorenz on a number of classical problems, including egg-rolling in geese; they showed that there was a fixed pattern of egg-rolling that was performed by geese of different species. Once the egg-rolling began, it continued even if the egg was removed from under the mother's beak. Tinbergen and Lorenz also showed that geese would choose to retrieve eggs that were orders of magnitude larger than their natural eggs; they called these eggs "supernormal" stimuli.

Tinbergen was a skilled and dedicated field biologist who studied a wide variety of behavior patterns ranging from homing in wasps to antipredatory behavior in birds. He stressed that in studies of behavior we need to pay attention to the evolution of the behavior, what caused it, how it helped animals to adapt to their environments, and how it developed. Ethology was truly an integrative science. Tinbergen was an eclectic biologist. Later in his life he and his wife studied human autism, applying ethological methods to learn about this disorder. Perhaps surprisingly, he refused to study the emotional lives or subjective states of animals because he believed we could never learn much about them.

Karl von Frisch was responsible for teasing apart what is frequently called "bee language." He discovered that through a variety of dances honeybees could tell other hive members where there was food and what kind of food was available. His research has stimulated many researchers to study cognition in bees, to determine how intelligent bees are, and to design clever experiments to determine if bees actually think about what they are doing.

The work of Lorenz, Tinbergen, and von Frisch played a large role in my own development as an ethologist. It was clear that many different points of view had to be incorporated into any study of animal behavior and that there was ample room for different sorts of approaches. Lorenz's sentimentalism and anthropomorphism were in marked contrast to Tinbergen's objectivity. Pluralism has been important to me as I delve into the lives of other animals, and I try to convey the importance of open-mindedness to my students and colleagues.

Playing with play

For the last thirty years I have been interested in the development of animal play—what animals do when they play, and why. To study the ways in which play unfolds in early life, one needs to be able to watch very young animals and follow known individuals as they grow up. Frequently, this is impossible to do in the wild because young animals are difficult to see, or they spend a great deal of time in their den or nest until they are ready to try to survive on their own. Basically, I would watch youngsters play, film their behavior, write down detailed notes on scoresheets, and then spend an enormous amount of time analyzing my notes and the films one frame at a time. It could take as much as five hours to analyze a mere twenty minutes of film. But these detailed analyses were necessary to learn about the nitty-gritty of play, how animals communicate their intentions to play and how they organize play with different partners. There is no substitute for this type of research, and while at times it truly was little fun, in the end, when I saw what I had learned, it was well worth it.

The geometry of grosbeak groups

I also used video analyses to learn about the ways in which western evening grosbeaks, remarkably beautiful and noble birds, detect potential predators such as ravens and magpies and avoid being eaten by them. In this study I was very interested in learning if there was a relationship between the number of birds in a group and how much time each bird devoted to scanning for predators—what is called "being vigilant"—and feeding. Previous researchers had discovered that, in many birds and mammals, the larger a group, the less time any given individual spent scanning and, as a result, the more time could be devoted to feeding. This was because there were "more eyes" to scan for danger.

I filmed birds for months on end and discovered that I had to repeatedly play back videotapes and study single frames taken every thirtieth of a second in order not to lose information about what the birds were doing. Then one day I had to go up on the roof of my house to retrieve a Frisbee for my companion dog, Jethro. When I was up on the roof I looked down at the birds and realized that a group of birds could be organized in various geometric arrays. They could be arranged in a line so that a single grosbeak could only see his or her nearest neighbors, or in a circle so that each bird could see every other bird. I felt stupid for overlooking this obvious fact, and it was an unplanned trip onto my roof that led to six more years of study. Serendipity surely paid off. I realized that perhaps the amount of

time individuals spent scanning or feeding was influenced not only by the
number of other individuals in a group but also by how they were arranged in
space. (I will tell you more about my results in chapter 3.) After about fifteen
years of studying animals, I learned something new that changed my whole
outlook on a long-term research project that had lasted for twelve years.

Penguins and their predators: Personal growth "on ice"

My research on Adélie penguins was an unbelievable experience. I lived in
an eight- foot by sixteen-foot hut with three colleagues at the Cape Crozier
penguin rookery on Ross Island, Antarctica. The work was grueling and
dangerous. But it was so much fun that the difficulties were lost in the pure
joy of seeing 250,000 penguins in the wild. A typical day in the field began
around five A.M. and often lasted until nine P.M., for the penguins did not
ask us when we would be available to watch them. We carried binoculars,
spotting scopes, and food, including a lot of chocolate in case we got lost or
the weather prevented us from returning to our hut. Our arsenal also included pens, pencils, waterproof paper, tape measures, stopwatches, and
crampons. We needed crampons so that we would not slide into the Ross
Sea while following the adroit penguins, who are amazingly adapted to living on ice and rocks and who are able to swim—actually, fly—through
freezing water frequented by unrelenting predators. The penguins would
spend a good deal of the day waddling here and there, stealing rocks from
one another's nests to build or reinforce their own, and squawking. They
also were curious about us and fearlessly approached to see what we were
doing. I once made the mistake of bending down to say hello and got swat-
ted by the edge of one penguin's wing and nipped in the behind by another's bill.

One aspect of our study involved territorial behavior in south polar
skuas, seagulls who prey on young penguins and injured adults. Skuas are
constantly looking out for intruders. I accidentally stepped into the territory of a big male skua, and four pounds of squawking gull came screaming
down on me, knocking me off my feet. I fell on my back, hit my head, and
crushed not only my field gear but eggs that we had collected in order to
study experimentally the response of gulls to penguin eggs. A season of
work went down the drain because I trespassed where I should not have
gone. The amazing thing was that when we mapped out the territory of
this particular male, we realized I had only stepped about two inches into
his kingdom.

The penguins showed little fear of us and allowed us to watch them.
How lucky we were. Penguins would often slide down to the water on what

we called "Penguin Highway." It was amazing to see how fast they would go with no fear. Penguins would line up on ice floes and wait for an individual to jump in to test the water for the presence of leopard seals or killer whales. After a few seconds, if the scapegoat surfaced, they all jumped in en masse, looking for krill and other food for themselves and their children. Often we would try to predict who would be the first to jump in the water—with little success, I might add. Sometimes we grieved as a badly mauled penguin surfaced, having encountered one of its large and efficient predators. We once saw a penguin emerge from the water, its chest torn open. Limping badly, it made it back to its chicks, where it lived for the duration of our study. These little penguins were incredibly durable and persistent, surviving encounters with huge predators. I felt ashamed for having been floored by a mere gull.

As a research experience, studying penguins and skuas was awesome. The time spent down "on the ice" provided for remarkable personal growth and was the beginning of my seeking answers to "big" questions that were not typically discussed in classrooms. My personal notes consistently contained thoughts questioning what types of research, including field studies of behavior and behavioral ecology, could be justified, what in the world I was doing in this pristine environment, and whether the knowledge that I was gaining was worth intruding on the lives of these amazing animals. These questions and others like them remain in my brain and my heart.

Onward . . .

Trying to answer different questions about different species requires that colleagues and I develop clever ways to ask animals how they live and to have them tell us in ways that are intelligible to us. Studying other animals is usually fun but can be tedious. It can also make one seem odd to those who do not do this sort of research.

We know a lot about the behavior of other animals, but there is so much more to learn. Although many people, perhaps most people, are aware that animals live in their own worlds and have their own ways of doing things, often animals are represented, or misrepresented, from only our point of view. In the next chapter I will discuss how animals are perceived and represented by humans. I will stress the importance of adopting an animal-centered perspective, rather than solely a human-centered, or anthropocentric, view. Although we have little choice but to sense the world through our own eyes, ears, and noses, our view does *not* have to exclude the animals' points of view.

two

REPRESENTING AND MISREPRESENTING ANIMALS

Elephants aren't couches

In spring 2001, Asian elephants were regularly moved in and out of the Denver (Colorado) Zoo. Dolly, a thirty-two-year-old female, was removed from her friends, Mimi, forty-two, and Candy, forty-nine, and sent to Missouri on her honeymoon, as the zoo called it, to breed. About the same time, Hope, a mature female, and Amigo, a two-and-a-half-year-old male who had been taken from his mother, were sent to the zoo. They lived next door to Mimi and Candy. Over time, Mimi became increasingly agitated, and in June she pushed Candy over and Candy had to be euthanized. Mimi was now alone. Two days after Candy died, and a day after she was autopsied within smelling distance of the other elephants, Hope escaped from her keepers and rampaged through the zoo. Luckily, no one was seriously injured. Hope was then transferred out of the zoo, and Rosie was brought in. Simply put, these intelligent and emotional beasts were moved about as if they were couches. It is known that elephants live in matriarchal groups in which social relationships are enduring and deep. Their memory is legendary. Elephants form lifelong relationships and grieve when bonds are broken because of separation or death. When elephants move in and out of groups, there can be severe disruption of the social order, and individuals can get very upset. This is what happened at the Denver Zoo.

Joyce Poole, who has studied African elephants for decades, shared her views with me: Having studied the nature and personalities of free-ranging elephants in Kenya for over twenty years, it is not surprising to me that

Hope, brought to Denver only three months ago, might be emotionally distraught just two days after Candy's death and autopsy. I once witnessed three young males spend an hour attempting to lift an unrelated female who had died before my eyes. Two days later I found the same three males standing solemnly over her body, touching over and over again her bloody face where people had hacked out her tusks. The scene still haunts me and is a reminder to me never to underestimate the depth of an elephant's understanding.

Animals as subjects and not objects

The words we use to talk about animals influence our attitudes about them—how we think about them, how we view them, and how we treat them. Many humans also organize their worlds into "in" groups and "out" groups. Members of out groups are viewed as others. While animals are certainly others, this does not mean that they are lesser beings or less valuable than their human kin. In his marvelous book *The Unbearable Lightness of Being*, the Nobel laureate Milan Kundera claims that the true moral test for mankind consists of its attitude towards those who are at its mercy: animals.

The magician and philosopher David Abram stresses, in his book *The Spell of the Sensuous*, that we live in a more-than-human world. If we compartmentalize animals into such distinct categories as "others" or "them," we lose much of the richness that makes us *all* animals. If we follow Aristotle, Augustine, and Thomas Aquinas in arguing for a hierarchical view of nature in which there are "lower" and "higher" animals, then we often make value judgments that portray higher animals as better or more valuable than lower animals, with lower animals existing solely for the service of humans. If we think that animals are merely robots or objects and refer to them as an "it" rather than a "who," then we will treat them as disposable commodities, much as we treat bicycles or backpacks. If we believe that animals are lower than humans, then we will treat them accordingly. Animals then become mere instruments for achieving human ends, and in some cases, such as when dolphins carry torpedoes, they become instruments of destruction. However, if we believe that animals are subjects of a life, and that they should be respected and treated with compassion, then we will interact with them with concern for their own views of their worlds and their feelings.

In this chapter I discuss how animals are perceived and represented and misrepresented in mass media and in other venues. I also describe how ethologists study behavior and dispel concerns (fears and myths) about anthropomorphism, the practice of attributing human characteristics to non-human animals (or to supernatural gods or inanimate nature). I also discuss the evolution of behavior and introduce the notion of speciesism.

Human perceptions of, and attitudes toward, animals

There are many concerns about how animals are represented in advertisements, on television, in movies, in books, in zoos, and in the news. Chimpanzees often are given human characteristics, especially the ability to talk. They also are dressed in clothes and are adorned with wigs and makeup. This practice is insulting and ignores the phenomenally interesting and mysterious world of chimpanzees.

The image of dolphins has also suffered at the hands of humans. Consider Flipper, the famous dolphin (actually there were five Flippers) whose image was created to serve human ends. Richard O'Barry, Flipper's first trainer, believed that Flipper was a fabrication of hundreds of people who created his legend. O'Barry's story, reported in his book *Behind the Dolphin Smile*, is a moving example of a man who came to realize that it was simply wrong to keep dolphins in tanks of water (often chlorinated) for commercial exploitation. The mentality that we can do whatever we want with such amazing animals as dolphins is what led to the exploitation of the image of Flipper and the development of numerous large and very profitable businesses, dolphinaria, each with its obligatory Flipper. In the United Kingdom alone it has been estimated that there have been more than a hundred deaths among the three hundred bottle-nose dolphins who were imported and kept in captivity.

Television is also a problem. Elizabeth Paul showed that television programs in Great Britain supported the notion of a hierarchy of "higher" and "lower" animals, with lower animals not suffering as much as higher animals. Cruelty to mammals was not tolerated, but cruelty to fish and invertebrates was acceptable. She also found that mammals tended not to be shown as meat for meals, and she believes that this is related to adults being uncomfortable with advocating kindness to animals and then allowing them to be killed for food. They do not want their children to know that a hamburger is a cow on a bun—and that the cow had a mother, father, and sisters and brothers. But some children know this. The late Francisco Varela told me that his young daughter said that she would not eat an animal because its mother would cry.

When an animal is shown in a setting that is unrelated to its natural environment, a false message is presented. This prevents communicating an accurate understanding of the animal's nature. TV commercials often include mountain lions or other animals to show that a sofa is soft and comfortable. Because the lion is shown out of context as a soft and cuddly creature, it does nothing to promote an understanding or appreciation of the true characteristics of lions, who they are and how they live. These distortions also convey a false picture of humans' place in nature.

The ways people view other animals often are related to how similar they are to humans, either in their behavior or their appearance, or how familiar people are with the animals they are discussing. People also often project their own perceptions onto animals, making them the beings they want to them to be. For example, many people see in their companion animals traits that they value, such as trust, devotion, and unconditional love. They then treat them accordingly. Great apes look like some of us, and as a consequence of this similarity they take on many of our characteristics. Scientists show different attitudes toward animals of the same species depending on whether they are encountered in the laboratory or at home.

Many people select certain species and use them as poster animals for various campaigns for and against them. For example, gray wolves and whales are charismatic creatures who find their way into the hearts of many people. In his book *The Value of Life: Biological Diversity and Human Society*, Stephen Kellert notes: "The whale in the sea, like the wolf on land, constituted not only a symbol of wildness but also a fulcrum for projecting attitudes of conquest and utilitarianism and, eventually, more contemporary perceptions of preservation and protection."

In discussions about the moral status of animals, there is a progressive trend for greater protection for wild and captive animals that might be due in part to an increasing number of people moving from farms and rural areas to more urban environments. This trend is very clear for marine mammals. In a survey of Americans' perception of marine mammals reported by Kellert, most respondents were opposed to commercial whaling, often for ethical reasons. Concern was also expressed about the commercial exploitation of seals, sea otters, walruses, and polar bears. Most Americans also objected to commercial whaling by native peoples or the resumption of killing gray whales. A majority of Alaskans opposed oil and gas development if it injured or killed marine mammals. Kellert found that a majority of Americans objected to the captive display of marine mammals in zoos and aquariums if there are no demonstrated educational and scientific benefits. They were concerned with the care given to captive individuals. (The unsuccessful legislative effort, prior to the reauthorization of the U.S. Marine Mammal Protection Act in 1988, to prohibit any invasive research involving marine mammals unless that research would directly benefit the subject of the research is another example of this kind of concern.) To date, there are no unambiguous data that show that there are any significant educational or scientific benefits for the animals, despite beliefs that such benefits accrue.

Kellert's study of American perceptions of marine mammals and their management shows that most people support the various goals of the Marine Mammal Protection Act. Most are willing to "render significant sacrifices to sustain and enhance marine mammal populations and species. . . .

These findings clearly indicate that marine mammals possess considerable aesthetic, scientific, and moral support among the great majority of Americans today." A recent study in Great Britain indicated that people are more willing to pay to conserve marine mammals than terrestrial mammals.

Kellert identified a typology of nine basic values that include the following attitudes: *utilitarian* (practical and material exploitation of nature; animals are valued for their use-value to humans), *naturalistic* (direct experience and exploration of nature), *aesthetic* (physical appeal and beauty of nature), *moralistic* (spiritual reverence and ethical concern for nature), *dominionistic* (mastery, physical control, dominance of nature), and *negativistic* (fear, aversion, alienation from nature).

The results of Kellert's carefully conducted studies indicate that many factors enter into people's attitudes toward animals, and they also show how ignorance and economics, along with cultural variations and gender differences, make for a complex potpourri of attitudes. Lack of education about the behavior and ecological role of animals is often the most important determinant in utilitarian, dominionistic, and negative attitudes toward animals. While many people do not take science into account, but rather act on their emotions and passions, science needs to be accessible and disseminate knowledge widely. Money is another key issue. Kellert quotes Peter Matthiessen: "The removal of financial incentives . . . makes conservationists of one and all."

In his comparisons of attitudes in the United States, Japan, and Germany, Kellert found considerable variation in all cultures but no real fundamental differences among them. All of the values appear in each culture, but they vary in content and intensity. Each provides a valuable lesson. From the West we get an understanding of the ecological connections that "bind life together in a vast matrix of interdependencies and relationships." A recent survey has shown that 70 percent to 90 percent of the general public who were questioned in Europe and in the United States "recognize the right of nature to exist even if not useful to humans in any way." The East offers values that center on kindness and compassion for animals and attempts at achieving "harmony and balance with the natural world." Tribal, or non-industrialized cultures show how "other organisms constitute parallel nations who, if one listens carefully and watches closely, can communicate a vast and enduring wisdom."

Porky and Phoenix: A pig and a calf

As I write this book, millions of animals across Europe are being slaughtered because of the threat of foot-and-mouth disease. In many interviews farm families cried passionately, not solely because of economic losses but also because the animals being killed were their friends, with names and personalities of their own.

Once an individual animal has been identified and named, there is an immediate change in the way he or she is perceived. This was shown dramatically during the terrible culling of animals enforced by the British government during the recent outbreak of foot-and-mouth disease. "Happiness as Porky Lives to Grunt Another Day," read a newspaper headline in April 2001. Porky is a Vietnamese potbellied pig who lived with a group of pensioners. His doting friends kept him inside their house, and the men sent to kill him were barred from entering. And then there was Phoenix, a white calf who was found, still alive beside her dead mother, five days after the herd was slaughtered. Phoenix was headline news for days, and the Ministry of Agriculture, responsible for the killing, was besieged with letters demanding that Phoenix be spared. Even Prime Minister Tony Blair added his voice to the people's demand.

In these cases, public pressure saved the lives of Porky and Phoenix. Millions of cattle and pigs were killed, and although people were horrified by the mounds of the dead, their hearts were not touched as they were by the plight of Porky and Phoenix, who were special because they were named and recognized as individual sentient beings. Most of the time the slaughter of animals for food goes on behind closed doors. We do not think of the individual beings who are killed so we can serve meat for our meals. We do not talk about eating cows or pigs, but beef and veal, pork and bacon. That is why stories like those of Porky and Phoenix are important: they remind us of the fact that animals are indeed individuals, just like us, each one with a name, each one with his or her own life filled with happiness and sadness, fear and despair.

Calling animals by name: Social bonds and science

My early scientific training as an undergraduate and a beginning graduate student was grounded in what the philosopher Bernard Rollin calls "the common sense of science," in which science is viewed as a fact-gathering value-free activity. Of course, science is not value-free—we all come to our lives with a point of view—but it took some time for me to come to this realization because of heavy indoctrination and arrogance concerning the need for scientific objectivity. In supposedly objective science, animals are not subjects but rather objects that should not be named or bonded closely with. However, naming and bonding with the animals I study is one way for me to respect them. Although some people believe that naming animals is a bad idea because named animals will be treated differently—usually less objectively—than numbered animals, others believe just the opposite, that naming animals is a good idea. Christopher Manes notes of many Western cultures, "If the world of our meaningful relationships is measured by the things we call by name, then our universe of meaning is rapidly shrinking. No culture has dispersed personal names as parsimoniously as ours . . . officially limiting personality to humans . . . [and] animals have become increasingly nameless. Some*thing* not some*body*."

In her book *Reason for Hope*, Jane Goodall, the world-famous expert on chimpanzee behavior, notes that early in her career she learned that naming animals and describing their personalities was taboo in science, but because she had not been to university she did not know this. She thought it was silly and paid no attention. Early in her career, Goodall had trouble convincing reviewers of one of her early papers that naming the chimpanzees she studied should be allowed. She refused to make the changes they suggested, including dropping names and referring to the animals as "it" rather than "he" or "she," or "which" rather than "who," but her paper was still published. Goodall opposed reductionistic, mechanistic science early in her career as she does now, and her bold and pioneering efforts have had much influence on developing scientists' views of animals as thinking and feeling beings.

Researchers working with nonhuman primates and cetaceans usually name the animals they study; we read about such great apes as Kanzi, Austin, Sherman, and Koko, or the dolphins Phoenix and Akeakamai, and we often see pictures of them with their proud human companions. We also read about Alex, an African gray parrot whom Irene Pepperberg has studied extensively. Most people do not seem to object to naming these individuals. Is it because the animals who are named have been shown to be smart and to have highly developed cognitive skills or deep emotional lives? Not necessarily, for these and other animals are often named before they are studied intensively. Or, in the case of most nonhuman primates, is naming permissible because these individuals are more similar to humans than are members of other species? Why is naming a rat or a lizard or a spider more off-putting than naming a primate or a dolphin or a parrot?

It is well known that increasing the distance between themselves and nonhuman animals is a common practice among both scientists and nonscientists. Among scientists, distancing might make it easier to use animals in research in which animals are harmed or killed. Among the devices used in distancing are objectifying animals by referring to them as "it" and "which" and using terms such as "collecting," "euthanizing," "sacrificing," and "culling" to refer to "killing."

I do not have any problem naming animals, and I know of no evidence that unequivocally shows that naming animals produces less reliable scientific results than referring to animals with numbers. Such primatologists as Jane Goodall, Roger and Debbi Fouts, Sue Savage-Rumbaugh, and Frans de Waal do not shy away from this practice, nor do researchers such as Cynthia Moss or Joyce Poole who study elephants. I wonder if some scientists name individuals when they study them and then drop the names when they write about them for scientific literature. In my own experience, I have found that naming an aggressive coyote Brutus or Harry does not mean I will come to see him as more combative than if I called him coyote A236.

The context in which animals are used can also inform attitudes that people have even to individuals of the same species. For example, scientists show different attitudes toward animals of the same species depending on whether they are encountered in the laboratory or at home; many scientists who name and praise the cognitive abilities of the companion animals with whom they share their home are likely to leave this sort of baggage at home when they enter their laboratories to do research with members of the same species. Based on a series of interviews with practicing scientists, the sociologist Mary Phillips reported that many of them construct "a distinct category of animal, the 'laboratory animal,' that contrasts with nameable animals (e.g., pets) across every salient dimension. . . . The cat or dog in the laboratory is perceived by researchers as ontologically different from the pet dog or cat at home." One answer to the question of why dogs (and other animals) are viewed differently at work and at home is that at work dogs are subjected to a wide variety of treatments that would be difficult to administer to one's companion.

Anecdotes and anthropomorphism

There are many different ways of describing what animals do. How one chooses to summarize what they see, hear, or smell depends on the questions in which one are interested. There is not only one correct way to describe or to explain what animals do or feel.

Anecdotes, or stories, always find their way into people's views of animals. Some of my colleagues dislike or ignore anecdotes because they are merely stories with little or no substance; they are not "hard data." Anecdotes are central to the study of behavior, as they are to much of science. As we accumulate more and more stories about behavior, we develop a solid data base that can be used to stimulate further empirical research, and yes, additional stories. The plural of anecdote is data. Stephen Jay Gould, the world famous paleogeologist, has stressed the importance of case studies in science. Anecdotes, like to anthropomorphism, can be used to make for better science, if we only let them.

Biocentric anthropomorphism: Humanizing animals with care

> We are obliged to acknowledge that *all psychic interpretation of animal behavior must be on the analogy of human experience.* . . . Whether we will or no, we must be anthropomorphic in the notions we form of what takes place in the mind of an animal.
>
> Margaret Washburn, *The Animal Mind*

Professor Washburn wrote the passage above in 1909. Unfortunately, many researchers have ignored what is so very obvious: we are humans and we have by necessity a human view of the world. The way we describe and explain the behavior of other animals is limited by the language we use to talk about things in general. By engaging in anthropomorphism we make other animals' worlds accessible to ourselves and to other human beings. By being anthropomorphic we can more readily understand and explain the emotions or feelings of other animals. But this is not to say that other animals are happy or sad in the *same* ways in which humans (or even other members of the same species) are happy or sad. Of course, I cannot be absolutely certain that Jethro is happy, sad, angry, upset, or in love, but these words serve to explain what he might be feeling. Merely referring to the firing of different neurons or to the activity of different muscles in the absence of behavioral information and context is insufficiently informative.

Using anthropomorphic language does not force us to discount the animals' point of view. Anthropomorphism allows other animals' behavior and emotions to be accessible to us. I believe that we can be *biocentrically anthropomorphic* and do rigorous science.

Frans de Waal, in his book *The Ape and the Sushi Master*, introduces the notion of anthropodenial, a practice in which a dualism, or distinct separation between humans and other animals, is suggested. Differences, rather than similarities or evolutionary continuity, are stressed. Charles Darwin repeatedly stressed that the differences among many species were differences in degree rather than differences in kind. He argued that differences in mental abilities, for example, were differences along a continuum. Here is a modern example that may clarify what Darwin meant: Rolls-Royces and less expensive Fords are both cars. The differences between Rolls-Royces and Fords are differences in degree—they are both cars—and not differences in kind. However, Rolls-Royces and motorcycles are different *kinds* of motor vehicles. Of course, animals are *not* objects, but we can still talk about differences in degree and differences in kind. Thus the differences in mental abilities between, for example, wolves and chimpanzees are differences in degree rather than differences in kind. This simply means that there are many similarities in the mental abilities of wolves and chimpanzees and that chimpanzees are not 100 percent different from wolves.

To make the use of anthropomorphism and anecdote more acceptable to those who feel uncomfortable describing animals with such words as "happy," "sad," "depressed," or "jealous," or those who do not think that mere stories about animals truly provide much useful information, Gordon Burghardt suggested the notion of "critical anthropomorphism," in which various sources of information are used to generate ideas that may be useful in future research. These sources include natural history, individuals' perceptions, intuitions, feelings, careful descriptions of behavior, identifying with the animal, optimization models, and previous studies.

Burghardt and others feel comfortable expanding science carefully to gain a better understanding of other animals. Some scientists are freely anthropomorphic outside of their laboratories. Bernard Rollin points out that some researchers feel very comfortable attributing human emotions to, for example, the companion animals with whom they share their homes. They tell stories of how happy Fido (a dog) is when they arrive at home, how sad Fido looks when they leave him at home or take away a chew bone, how Fido misses his buddies, or how smart Fido is for figuring out how to get around an obstacle. Yet when the same scientists enter their laboratories, dogs (and other animals) become objects, and talking about their emotional lives or how intelligent they are is taboo.

Anthropomorphism as a disease

In my own studies I am carefully anthropomorphic, and I wonder why some of my colleagues view anthropomorphism as a disease. For example, John S. Kennedy, in his scathing dismissal of anthropomorphism in his book *The New Anthropomorphism*, writes that we can be confident that anthropomorphism will be "brought under control, even if it cannot be cured completely. Although it is probably programmed into us genetically as well as being inoculated culturally that does not mean the disease is untreatable." Kennedy also claims that "anthropomorphism must take its slice of the blame for a sort of malaise that has lately afflicted the subject of ethology as a whole." The idea that there is a malaise in the field of ethology simply does not correspond with current interest in the field. Indeed, there seems to be unprecedented interest in animal behavior worldwide, and much of this appeal is initially stimulated by using human terms to discuss nonhuman animals.

Kennedy is typical of critics who write as if the only alternatives are an unconstrained use of anthropomorphism on the one hand and the total elimination of anthropomorphism on the other. But there is a middle position. Anthropomorphism can be useful if it serves to focus attention on questions about animal behavior that might otherwise be ignored. Anthropomorphism might be used in a rigorous way to help in theory construction and to stimulate empirical research projects. Carolyn Ristau notes that in her attempts to study injury-feigning in plovers under field conditions, the cognitive ethological perspective that entails some anthropomorphic thinking "led me to design experiments that I had not otherwise thought to do, that no one else had done, and that revealed complexities in the behavior of the piping plover's distraction display not heretofore appreciated." Being anthropomorphic does not ignore the animals' perspectives. Rather, anthropomorphism can help to make accessible to us the behavior and thoughts and feelings of the animals with whom we are sharing a particular experience.

Field studies "versus" captive studies

Many people believe that experimental research in such areas as neurobiology constitutes more reliable work and generates more useful (hard) data than, say, ethological studies in which animals are "merely" observed. However, research that reduces and minimizes animal behavior and animal emotions to neural firings, muscle movements, and hormonal effects will not likely lead us significantly closer to a better understanding or appreciation of the activities or feelings of animals. For example, concluding that we will know most, if not all, of what we can ever learn about animal emotions when we have figured out the neural circuitry or hormonal bases of specific emotions will produce incomplete and misleading views concerning the true nature of animal (and human) emotions.

All research involves leaps of faith from available data to the conclusions we draw when trying to understand the complexities of animal emotions, and each has its benefits and shortcomings. Often studies of the behavior of captive animals and neurobiological research are so controlled as to produce misleading results concerning social behavior and emotions because animals are being studied in artificial and impoverished social and physical environments. Experimenters themselves might put individuals in thoroughly unnatural situations. De Waal notes that "certain social phenomena cannot be transferred to the laboratory," because it would be impossible to recreate the social and nonsocial environments that are responsible for producing and maintaining ongoing encounters, such as those observed in dominance interactions. William McGrew, who has studied chimpanzee behavior for decades, goes a bit further and stresses that "no experiment has ever simulated any of the delicate probing tasks such as termite-fishing shown by wild chimpanzees." Some researchers have discovered that many laboratory animals are so stressed from living in captivity that data on emotions, other aspects of behavior, and physiology are compromised from the start.

There is even much concern about such common laboratory animals as white mice and white rats, who are routinely used in a wide variety of behavioral research projects. It turns out that individuals raised and maintained in sterile conditions of captivity with little enrichment show signs of boredom and stress when they are not being studied. When researchers used infrared cameras to spy on animals at night when they were not being observed by the experimenters, mice and rats who behaved normally when they were being observed directly showed what are called cage stereotypies—repeated cage biting and cage scratching—indicative of bored and stressed animals. (Bears and many other zoo animals perform stereotyped pacing, walking back and forth in highly predictable patterns.) There is evidence that these deprived rodents may also suffer from damage to the basal ganglia, a part of the brain that regulates the initiation of movement. Rodents kept in enriched environments perform better on memory tests than rodents—perhaps brain-damaged animals—kept in sterile cages. Researchers caution that they might actually be producing unreliable data if they wish to generalize about the behavior of wild animals from the behavior of impaired captive relatives. These animals might not be good models of normal behavior. I will return to this topic in chapter 8.

Field work also can be problematic. It can be too uncontrolled to allow for reliable conclusions to be drawn. It is difficult to follow known individuals, and much of what they do cannot be seen. However, it is possible to fit free-ranging animals with telemetric devices that transmit information on individual identity, heart rate, body temperature, and eye movements as the animals go about their daily activities. This information is helping researchers to learn more about the close relationship between physiology and behavior.

There are better and worse places in which to study behavior. The setting in which a behavior pattern is studied needs to be determined by the questions of interest. Only careful observations under field *and* captive conditions will permit us to assess just what the important variables are that influence how individuals interact with their social and nonsocial worlds, how these different variables are used, and how they may be combined with one another.

Describing and categorizing behavior

To appreciate the behavior and lives of animals, it is important to know what they do, to describe the various activities in which they partake during the course of an average day and throughout the year. Here I discuss various aspects of how to study animal behavior. I hope this brief discussion helps you understand the ways in which ethologists go about their business and serves as a guide if you choose to conduct your own studies. No matter how informal they may be, I guarantee that your studies will be fun and a learning experience to boot.

Many people mistakenly believe that it is easy to study animal behavior. Those who make a living studying animal behavior already know that studies of behavior are rather difficult. Wild animals do not perform on cue for researchers. Patience is needed when you consider that many animals basically rest and "do nothing" about 90 percent of the time. It also takes time for animals to habituate to the presence of researchers. When we studied coyotes in the Grand Teton National Park, it took a few months before they continued doing what they were doing when we arrived at the field site. Because this was a relatively unexploited population, at first the coyotes were very curious and wary, and they stopped what they were doing and stared at us. They also kept their distance. Ellen van Krunkelsven and her colleagues discovered that the best way to habituate wild bonobos in Congo to the presence of observers was to approach them in a nonthreatening way by being silent and unobtrusive and making their presence known only when sitting down quietly.

Behavioral studies usually start with the observation and categorization of behavior patterns that animals perform. Despite claims of untrained observers that animals perform only a handful of actions, dogs and other animals have rather large behavioral repertoires. I discovered that the number of different actions that can be recognized in a single study of coyote behavior can easily be fifty or more. Stuart Altmann listed more than 120 behavior patterns for rhesus monkeys.

Observation and description usually lead to the development of an ethogram, or behavioral catalogue, of the actions that have been tallied. Descriptions included in this menu can be based on visual information (what an action looks like), auditory characteristics of sounds (frequency, duration, or amplitude), or chemical constituents of glandular deposits, urine, or feces. Great care needs to be given to the development of an ethogram, for it is an inventory that anyone else should be able to replicate without error. Thus, for example, if one calls an action a "bow" when a dog crouches on her forelimbs and invites play (see chapter 6), then others interested in this pattern would need to know what it looks like and not mistake a bow for another action, such as stretching. Not all movements are behavior patterns to be categorized. One would not, for example, list the movement of an armadillo dragged fifty meters along the highway by a speeding pickup truck. For most studies, it turns out that after a little training, many actions are easily recognizable, and reliability across studies is not a major problem.

A good ethologist asks what it is like to be the animal under study and tries to develop an awareness of the senses that the animals use singly or in combination with one another. It is highly unlikely that individuals of any other species sense the world the same way we do, and it is unlikely that even members of the same species sense the world identically all the time.

Female and male perspectives

The scientific vocabulary that is used to describe behavior varies among researchers and even between male and female observers. Elizabeth Adams and G. W. Burnett discovered that female ethologists working in East Africa use a substantially different descriptive vocabulary than do male ethologists. Of the nine variables they studied, those concerning cooperation and female gender were the most important discriminating women's and men's word use. They concluded: "The variable COOPERATION demonstrates the appropriateness of feminist claims to connection and cooperation as women's models for behaviour, as divergent from the traditional competitive model." Why women and men approach the same subject from a different perspective remains largely unanswered.

Description, explanation, and speciesism

The term "speciesism" was originally coined by the British psychologist turned animal advocate Richard Ryder. In the context of animal use, speciesists make decisions about how humans are permitted to treat other animals based on an individual's species membership rather than on that individual's characteristics. Thus, for example, all and only humans or all and only mammals might constitute protected groups. Recently, the British and New Zealand governments declared a ban on the use of great apes in research. This precedent-setting decision is speciesist. It was been argued that these primates deserve special treatment because of their cognitive capacities.

Nonspeciesists, in contrast to speciesists, use individual characteristics to make ethical decisions about animal use. Careful attention is given to individual variations in behavior within a species.

Speciesists also often use such words as "higher" and "lower" to refer to different groups of animals. However, using such words and ranking species by drawing lines to place different groups of animals above and below others are extremely misleading because they fail to take into account

SPECIESISM	NON-SPECIESISM

Humans (H)

Gorillas (G)

Chimpanzees (Ch)

Monkeys (M)

Dogs (D)

Cats (C)

Birds (B)

Fish (F)

H_1 H_2 H_3 H_4

G_1 G_2 G_3 G_4

Ch_1 Ch_2 Ch_3 Ch_4

M_1 M_2 M_3 M_4

D_1 D_2 D_3 D_4

C_1 C_2 C_3 C_4

B_1 B_2 B_3 B_4

F_1 F_2 F_3 F_4

A representation of speciesist and nonspeciesist perspectives. The speciesist view (left) shows eight species chosen for illustration. Lines dividing them into a linear hierarchy suggest, for example, that humans are "higher than" gorillas and chimpanzees and that monkeys are "higher than" dogs and cats. Speciesism provides a convenient way to make difficult decisions about which species may be used in different types of human activities, including research and teaching. However, this view does not pay attention to evolutionary continuity, and it deemphasizes individual variability. The nonspeciesist view (right) stresses individual variability, even within species. Lines encircling different individuals (H1 and G1; D2 and C1) illustrate that individual differences are important to consider. That is, it is possible that individual members of different species may be "equivalent" with respect to various traits or that individuals of a given species may possess characteristics that are exclusively theirs. Also, individuals of species that are thought to be "lower" than others may be more skilled in certain areas or experience pain, anxiety, and suffering more than individuals of species that are thought to be "higher." Nonspeciesist views argue against the use of species membership as the sole criterion for choosing which individuals should be used in various types of human activities.

the lives and worlds of the individuals themselves. Speciesism also can ignore evolutionary continuity. Deciding which among the criteria that are used to place species in some hierarchical order are morally relevant, and how evaluation of these criteria is to be made, would present serious problems even if one were able to argue convincingly for the use of a single scale.

Speciesists often use taxonomic or behavioral closeness to humans, similar appearance, or the possession of various cognitive capacities displayed by normal adult humans to draw the line that separates humans from other animals. Cognitive abilities include the capacities to be self-conscious, to engage in purposive behavior, to communicate using a language, to make moral judgments, and to reason (rationality). Using these criteria, most animals cannot qualify for protection. But there are some humans (young infants and, unfortunately, senile adults) who cannot qualify either, and this can be a problem for speciesists who rely on cognitive capacities. Because of individual differences within a species, this view from the top, a human-centered "them" versus "us" perspective, can be difficult to apply consistently.

In practice, when deciding about the types of treatment to which animals can be exposed, speciesism often is narrowly used to mean "primato-centrism" or "humanism," and human superiority is often implied in speciesist arguments. However, individuals representing many other species experience pain and suffering (physically and psychologically), even if these are not the same sorts of pain and suffering that are experienced by humans, or even other animals, including members of the same species.

An argument against "we/them" dualisms

"We" versus "them" dualisms do not work. It is the similarities rather than the differences between humans and other animals that drives much research in which animals' lives are compromised. If "they" who are used in research are so much like "us," then much more work needs to be done to justify certain research practices. Although we are very different from other animals, is it truly believable that humans are the only species that can think, feel pain, experience anxiety, and suffer? Even if we are very different from dogs or cats, there is no reason to think that dogs, cats, and many other animals do not think in their own ways and do not feel pain and suffer in their own ways.

Portraying animals as objects promotes the view of animals as commodities. It is essential that we instead try as hard as we can to answer the question "What is it like to be a _____ ?" When you watch animals make every effort to take their point of view. Try to "mind animals." Imagine what their worlds are like to them. What is it like to be a bat, flying around, resting upside down, and having very sensitive hearing? Or what is it like to be a dog with a very sensitive nose and ears? Imagine what it is like to be a free-running gazelle, or a wolf, coyote, or deer out in nature. Take advantage of what animals offer to us. It is essential not only to look at animals but to see them as they actually are, not as we want them to be.

Studying the behavior of other animals should also be fun. David Macdonald, who has studied the behavior of red foxes and many other animals around the world for almost three decades, still loves to go out into the field and watch them. In his book *Running with the Fox* he exclaimed: "I study foxes because I am still awed by their extraordinary beauty, because they outwit me, because they keep the wind and the rain on my face . . . because it's fun."

three

◘

THE RICHNESS OF
BEHAVIORAL DIVERSITY

A Potpourri of Animal Behavior

When playing dumb is being smart

People are continually amazed at the social, learning, and cognitive skills of many animals. Animals are said to be smart when they perform such tasks as counting objects, forming concepts in which differences or similarities are recognized, making and using tools, deceiving others, or using complex forms of communication.

Animals might know what others know, but sometimes it is smart to play dumb. Are animals smart enough to play dumb? Can they appraise their social situation and change their behavior depending on who is around? Yes. Wolves store and retrieve food more often when others are not looking. Rhesus monkeys will not emit food calls indicating that food is available if they are in the presence of other monkeys. And chimpanzees often ignore food, thus leading researchers to think they are dumb, only to retrieve the food when other group members are not around. It is too simplistic and anthropocentrically arrogant to assume that animals other than humans do not control their behavior according to who is watching.

Recently, Christine Drea and Kim Wallen discovered that low-ranking rhesus monkeys will play dumb in certain social situations. Drea and Wallen studied monkeys as they learned to discriminate boxes that contained food from those that did not. They compared the performance of monkeys tested in the presence of all members of their social group with their performance in groups of only more dominant or only more subordinate monkeys. They then reversed the situation and tested monkeys on the same problem; that is, a monkey previously tested in the company of only dominant individuals was then tested in the company of only subordinate monkeys, and vice versa.

The results of this creative study are very interesting. Dominant monkeys performed well in all conditions but subordinate monkeys performed well only when they were apart from higher-ranking animals. Because all monkeys had previously learned the task, Drea and Wallen concluded that the subordinate monkeys were indeed playing dumb—they were voluntarily inhibiting their behavior depending on who was around. Subordinates who learned the discrimination when alone showed a performance decline when intimidating higher-ranking animals were nearby; a subordinate individual's status relative to other monkeys made it advantageous to play dumb.

Social context clearly influenced performance on a previously learned task. The presence of dominant monkeys suppressed the expression of knowledge by subordinate animals, and monkeys of different social classes were influenced differently. This study opens up the door for future field studies on a wider array of species, including humans. There is also a strong message for those interested in human learning and the expression of scholastic accomplishments.

For humans it is known that social status, gender, and racial differences can lead competent individuals to inhibit academic or athletic performance. These findings stress that learning must be studied in the absence of intimidation, because humans, monkeys, and perhaps individuals of other species inhibit themselves when they are in uncomfortable social situations. Had monkeys only been studied in the presence of dominant individuals, Drea and Wallen might have concluded that subordinate individuals were dumber than dominant animals, not that they were simply playing dumb for good reasons.

Low-ranking monkeys, when playing dumb, are really playing it smart; they decide when to let others know what they know! And so do humans, often to individuals' detriment when they withhold knowledge because they fear retribution from potential bullies. What they do isn't an expression of what they actually know. Certainly, the long-term consequences of withholding knowledge could be very disadvantageous in an educational system that does not factor in the context in which performance is assessed.

Sperm wars

"Your daddy ain't your daddy, but your daddy don't know." So sang the Kingston Trio back in the 1960s. If only they knew what we know now.

In many species, sperm compete with one another. Males who think they are fathers are not. Multiple paternity occurs when more than one male's sperm produces individuals in a single litter. It has been discovered in various ground squirrels and in black bears, gray seals, domestic dogs, and many birds and insects. In Gunnison's prairie dogs, as many as one-third of all litters are multiply sired.

Sperm do not actually duke it out. Sperm competition is inferred from techniques such as DNA fingerprinting that indicate multiple paternity. Evolution has produced various adaptations that support explanations that invoke sperm competition. These traits—mate-guarding, multiple matings and ejaculations, the use of plugs to seal females' reproductive tracts—lessen the chance of more than one male mating with a female or of a future male's sperm being used. They also reduce the possibility of old sperm being used.

In many insects, later breeders usually are more successful in inseminating females than early breeders. In damselflies, males use their penis to scoop out sperm from previous males who have mated with a female. In dunnocks, sparrow-like birds, males who mate first peck at the female's cloaca, causing her to eject other males' sperm.

Females may play a direct role in controlling paternity. In barn swallows, females paired to short-tailed males seek extra-pair copulations ("affairs") with more attractive neighbors with longer tails. In black-capped chickadees, zebra finches, and indigo buntings, females prefer high-quality males for affairs. Sperm competition in animals seems to favor attractive males who are large and long-lived. In guppies, sand lizards, some rodents, and blue tits, multiple matings by females produce more viable offspring.

Why should females mate with more than one male? There is little evidence that they acquire higher-quality sperm, acquire more sperm, or have higher chances of having all their eggs fertilized. However, in dunnocks, Galapagos hawks, and marmosets, multiple matings result in more male care to offspring, which is correlated with increased survival of youngsters.

What about human sperm wars? There is certainly no dearth of interesting theories. For example, males produce a lot of deformed sperm, and it is speculated that these "kamikaze" sperm form a copulatory plug in the female's cervical canal. It has also been hypothesized that masturbation gets rid of old suboptimal sperm, leaving sperm more likely to result in conception. Females may be able to manipulate the likelihood of conception by controlling their orgasms. (Orgasms may help to transport the content of the upper vagina into the cervix). Female Japanese macaques experience more orgasms when they mate with high-ranking males than with low-ranking males.

While multiple matings within the life span of sperm and ova seem to be uncommon, a survey estimated that the conception of about 4 percent to

12 percent of children born in Britain in the late 1980s occurred while their mother's reproductive tract contained more than one male's sperm. Indeed, it is thought that sperm can hide in cervical crypts and retain fertilizing capacity for up to two weeks. Perhaps the Kingston Trio was more correct than they knew.

Researchers in biology, behavioral ecology, psychology, sociology, anthropology, and law want to know why sperm wars occur in some species and not others, what the benefits and costs of multiple matings are, how mates are chosen, what features females favor for extra-pair copulations, what "attractive" means, and how much males and females can control fertility. There is still a lot of mystery concerning sperm competition and who controls who, what, when, where, and why.

Dirty birds

Male rock ptarmigan who live in the Canadian tundra change color after they have found a mate. In the winter, males are white and blend in with snow. As the snow melts and they are no longer camouflaged, they suffer high predation. However, until a male forms a relationship with a female, he remains white. His white plumage seems to tell a female that he is able to avoid predators and is a good choice as a mate. But soon after his mate is incubating eggs and can no longer be fertilized, the male rolls in mud and water. The result of the soiling of their plumage is to allow the males to blend in with the rocks where they live and thus be less conspicuous to predators. Rolling in mud is quicker than molting and growing new feathers. Rolling also allows males to be flexible. In one case, a female lost her clutch of eggs and became sexually receptive. Her mate then cleaned himself up so as to remain attractive. Robert Montgomerie, at the University of Ontario in Canada, notes that he and his colleagues can spot a white male from as much as a kilometer away but walk right by them when they are dirty. Another interesting observation is that this is the first time a bird has been observed to make himself dirty in order to disguise himself. Perhaps this is an evolutionary precursor to human males, who when they begin courting a female do their best to look spiffy but soon after give less concern to their appearance.

Watching and enjoying animals

There is no better way to come to understand and appreciate the richness of animal behavior than to watch animals. However, most people are not able to do this in the wild, so videos, zoos, and magazines have to suffice. Although I make my living studying animal behavior and reading about others' work, I am constantly amazed at what my colleagues and I discover about the often mysterious ways of our animal kin. The following examples, many of which stem from my own longtime interest in animal behavior, form a bridge to the "big" topics to which I will turn in later chapters.

What is it like to be a _____ ? Taking the animal's point of view

What is it like to be a given animal? Often, observations and experimental results make little sense until the animal's perspective is taken into account. Why do some baby birds kill their siblings? Why do adults sometimes kill others' youngsters? Why do males often put so much time and energy into courting females that they have little time or energy for anything else? Why are some animals picky when it comes to food? Why do birds fly in V formation? Do animals gossip? Do animals have a sense of humor? How do birds decide when to feed and when to look for predators, and what factors influence their choices? Do sperm actually compete? Is there an optimal ratio for waist to hip size that influences a female's attractiveness to males? Why do some animals help to raise children who are not their own? What is so special about "yellow snow"? Do animals have culture? Can animals be their own pharmacists? Is homosexuality prevalent in animals? Do animals dream, and if so, what do they dream about?

E. O. Wilson's "sociobiology express": Surrounded but not defeated

In 1975, E. O. Wilson's difficult-to-tote tome *Sociobiology: The New Synthesis* appeared. One of the century's most important books in behavioral biology, this volume was, and remains, a gold mine of ideas. Wilson defined sociobiology as "the systematic study of the biological basis of *all* [my emphasis] social behavior." Sociobiology, along with behavioral ecology, was destined to consume ethology. Wilson wrote: "The future, it seems clear, cannot be with the ad hoc terminology, crude models, and curve-fitting that characterize most of contemporary ethology and comparative psychology." This is interesting because so much of behavioral ecology is indeed distinguished by these three characteristics, and ethology and behavioral ecology still have rather bright futures almost thirty years post-*Sociobiology*. Indeed, behavioral ecologists and sociobiologists always need to know nitty-gritty details about behavior, information that is provided by ethologists! Surely there is room for all sorts of behavioral science.

Many people jumped on the sociobiology bandwagon, whereas others, myself included, saw much promise in the sociobiological approach but also recognized that perhaps some of the flexibility in the social systems of various animals could be studied and understood without resorting to the vise of overbearing genetic constraints. While genes are necessary, alone they are not sufficient to explain a wide variety of behavior patterns, including within-species variations that arise due to differences in such ecological conditions as food availability and social factors including group size and group composition. Coyotes, for example, show great variation in group size depending on whether there is enough food to support a pack or only a pair of individuals. Many other species show similar variability. In addition, in situations where there is a pack of coyotes, some individuals help to rear others' offspring. This does not occur when only the mated pair live together.

Wilson's ideas were very challenging because they established an arena in which to exchange ideas concerning possible relationships between behavioral flexibility and genetic constraints. However, there was much resistance to sociobiology because of political overtones concerning the role of genetic constraints on human behavior that would not allow for much flexibility due to varying or changing environments.

Historical accounts of behavior: The privacy of evolution

Much research on the evolution of behavior has been stimulated by the seminal work of William D. Hamilton, who died prematurely in 2000. Hamilton's early papers, published in 1964, began a "revolution" in evolutionary studies of behavior. Hamilton, like Darwin, was especially interested in the evolution of altruism, in which one individual suffers a loss of reproductive fitness when it provides aid to another. Hamilton stressed the importance of "kin selection" in evolution, a process that results in relatives being given preferential treatment over nonrelatives. Thus an individual may prefer to provide food to brothers or sisters rather than to nonrelatives because his siblings share with him a higher percentage of genes than do unrelated individuals.

While evolutionary explanations are very helpful in understanding why we see certain patterns of behavior in extant animals, evolutionary accounts can also give a false sense of what we truly know and understand. Many behavior patterns have complex origins and persist in a behavioral repertoire because of biological and other factors (for example, psychological or sociological) or "causes."

The philosopher Elliott Sober differentiates between what he calls "selection for" and "selection of" characters. It is important to mark this distinction when offering evolutionary explanations. According to Sober, "selection for" means that a particular trait was directly or *causally* selected in evolution, whereas "selection of" means that a trait exists because of its association or correlation with other traits. Thus long legs and long fur might be associated with being able to run fast, but only long legs are causally related to this ability. Individuals with long legs also happen to have long fur, but long fur does not influence running speed. In Sober's explanatory scheme, there was "selection for" long legs and "selection of" long fur. Often it is difficult to tease apart which variables directly influence a trait and which are there as "hitchhikers."

It also is very difficult to reconstruct the past. Because behavior does not fossilize as do anatomical structures, the process of evolution often is a private matter that can at best only be hinted at. While there are limits to our knowledge of evolution, some evolutionary explanations are better than others, and we have learned much about the evolution of a wide variety of behavior patterns as a result of very careful studies, open minds, and interdisciplinary discussions. Similar problems plague people who attempt to study animal minds, because others' minds are private matters.

As a biologist, I approach my research in an evolutionary and ecological framework. My research integrates conceptual, theoretical, and empirical work and stresses that behavioral responses are adaptations (the evolutionary view) to varying environmental conditions (the ecological framework). I try to follow the guidelines that Niko Tinbergen laid out for the study of animal behavior. Tinbergen stressed that in studies of behavior we need to pay attention to the *evolution* of the behavior, what *causes* it, how it helps animals *adapt* to their environments, and how it *develops*. The essential part of many studies of the evolution of behavior, following the lead of Charles Darwin, is to gain an understanding of how the performance of a given behavior influences an individual's reproductive behavior, his or her reproductive fitness. It also is important to gain an understanding of how animals adapt to the environments in which they live from day to day.

Social bonding and dispersal in coyotes and red foxes

I have been interested in studying the development of behavior to learn how individual differences arise. I discovered that wolves, coyotes, and dogs show differences in social play behavior and aggression when they are as young as fourteen days of age. I wanted to know why coyotes (and golden jackals and red foxes) engage in intense fighting when they are three to four weeks of age, the result of which is a dominance hierarchy that persists at least until the animals are six months old, whereas young wolves and most dogs do not engage in such intense fighting. I was also interested in learning how individual differences in development are related to dispersal. (Dispersal occurs when animals leave the place where they were born and try to make it on their own.) These interests, in turn, led me to try to understand within-species variation in behavior and social organization. Why, for example, do coyotes living in different areas show variations in social organization such that in some habitats coyotes live alone, whereas in others they live with a mate or in large family groups, or packs? By studying within-species variation it became very obvious that speaking about *the* coyote was misleading, for coyotes, like many other species, show variations in behavior from one locale to another.

I hypothesized that social bonding, affiliative behaviors such as social play, and dispersal were related, and that individuals who were unable to form close social bonds with littermates, either because they avoided them so as not to be beaten up (there are indeed scapegoats among animals), or because they were avoided because they were "bullies," would eventually leave their group without being forced out. After years of collecting data on individual patterns of development in coyotes, my co-workers and I concluded that individual coyotes usually left their natal group of their own accord and that they were not forced out by aggressive siblings or parents. For some unknown reasons they decided that they would be better off on their own.

The importance of social bonding and affiliative behavior in dispersal has recently been demonstrated in urban red foxes for males but not females. While there have been no direct field tests of relationships between individual social development and later patterns of dispersal, Stephen Harris and Piran White devised an alternative and very novel means of assessing this relationship in red foxes. In red foxes, ear tags are chewed during social grooming. Harris and White studied patterns of grooming by looking at teeth marks left on ear tags and found that for males there was a strong negative relationship between grooming (an affiliative behavior) among littermates and future dispersal. Their data supported some of the ideas I had put forth in a paper about individual development and dispersal, namely that individuals who are less tightly bonded to their group will be more likely to leave their littermates.

I use these examples to show how field research combined with investigations on captive animals can produce important results that could not be obtained in only one or the other situation. Detailed developmental data on individually identified coyotes and foxes (and many other carnivores) are nearly impossible to collect under field conditions. Thus, although we could not be absolutely certain that a particular individual was the dominant member of its litter, based on patterns of social interaction, especially play solicitation, avoidance, and the initiation, escalation, and termination of fights, we felt comfortable assigning different dominance ranks to individual coyotes. For example, knowing that dominant coyotes were avoided in captivity and that subordinate coyotes avoided interaction in captivity allowed us to assign dominance rank to many individuals. Furthermore, knowing that dominant coyotes engaged in self-handicapping and role-reversing in play, and knowing that other animals were reluctant to play with dominant coyotes if the dominant animal did not previously signal an intention to play, allowed us to make reliable judgments about social rank among littermates.

When selection pressures disappear: Remembrance of things past

Often the ancestral selection pressures that were responsible for the evolution of a given behavior continue to persist. However, it is also possible that particular behavior patterns that we see today are relics of things past. This appears to be the situation for antipredatory responses to carnivores shown by such naive prey animals as American pronghorn, elk, and moose, species in which individuals have not been exposed recently to specific predators. Nonetheless, individuals still fear past predators who have not been around for a long time. For example, John Byers, at the University of Idaho, discovered that antipredatory strategies of American pronghorn make little sense given that there are no large predators who presently share their range. But pronghorn still display antipredator behavior that would have been appropriate when they coexisted with such large predators as saber-tooth tigers and cheetahs long ago. Pronghorn continue to gather in large groups and seek safety in large numbers even in the absence of fearsome predators. The historical perspective helped to solve the mystery of present-day pronghorn behavior.

Individual differences in antipredator behavior

Joel Berger's research is nicely related to John Byers's work because it is concerned with understanding antipredator behavior from an evolutionary perspective. I am proud to say that they were my first (Joel) and second (John) Ph.D. students at the University of Colorado.

    In many parts of the world, carnivorous predators such as wolves and brown bears have been eliminated due to exploitation by humans. Thus there are prey species that have not had recent exposure to these predators. But recolonization and reintroduction programs have now reexposed animals to predators that had been locally extinct. Berger wanted to know if these prey would be susceptible to predation by unfamiliar predators. He and his colleagues studied moose in Alaska, Wyoming, Sweden, and Norway in areas with predators, without predators, and where predators have recently recolonized. They hypothesized that predator-naive individuals may be less sensitive to cues provided by dangerous predators and studied the following behavior patterns: how vigilant or wary moose were, how often they behaved aggressively toward predators, and how often adult female moose left feeding sites after they were exposed to unfamiliar and familiar auditory (sound) and olfactory (odor) cues.

Berger discovered that predator-naive moose were much less responsive than other moose to auditory and olfactory cues. Moose who were subject to predation by wolves showed an increase in vigilance of 250 percent when exposed to wolf calls. They were also about six times more responsive to calls of ravens than were predator-naive moose, most likely because scavenging ravens are associated with bears and wolves. Naive moose never were observed to leave a feeding site in response to auditory or olfactory cues. They just kept on feeding.

Naive moose were also about one-fourth as likely to respond aggressively to odor cues as predatory-savvy moose. They did not lower their heads or retract their ears, and the fur on the nape of their neck did not stand on end—all responses that are shown when moose fight off bears or wolves. Naive moose also actually approached odors in about 16 percent of the trials, whereas moose who had experience with predators never did so.

All in all, individuals from populations that had not been recently exposed to predators are more vulnerable to predation. In a short period of time they have lost their protective responses. But do these predator-naive moose actually suffer increased mortality? Berger and his colleagues studied this question by analyzing the efficiency of bear predation on naive and experienced moose. They discovered that naive moose not only lack astuteness to predators but also experience what they called a "blitzkrieg." Clearly, what you do not know will harm you.

How do naive prey avoid extinction? One possibility involves the rapid development of predator recognition based on individual experience. Indeed, one model showed that predator-naive moose mothers whose calves were killed by wolves who recolonized Jackson Hole, Wyoming, after being reintroduced into Yellowstone National Park, showed responses increased by as much as 500 percent to wolf calls and were much less likely to resume feeding; the latency to return to feeding increased from a mean of less than thirty seconds to more than six minutes. It is interesting to note that moose mothers whose young died of starvation or due to being struck by cars displayed lower vigilance than mothers who had experienced wolves. Thus mothers who have lost offspring to predators are hypersensitive, and this might contribute to increased survival of their young.

Berger's study is an example of very clever experimental field work that tests important hypotheses. Naive prey can indeed rapidly develop the ability to process information about predators and avoid them. Moose do it in a single generation. Because naive prey are being exposed to predators as efforts are made to conserve large predators, this is a very useful adaptation and is good news for reintroduction programs. Predators are being reintroduced in about 175 locations around the world.

The next three stories are concerned with research dealing with social communication and what individuals might know about what others know. There is much interest in whether animals have a "theory of mind"— whether they know what others are thinking or feeling.

Do animals gossip? Grooming, big brains, and language

Do nonhuman animals really gossip? "Wow, did you hear that Mary killed a baboon all by herself?" "Hey, have you heard that Joe asked Helen to attend the Chimpanzee Ball?" Recent evidence suggests they might.

In his book *Grooming, Gossip, and the Evolution of Language*, Robin Dunbar, a British primatologist, claims that "being human is all about talking" and that there are evolutionary precursors to human gossip.

Much of Dunbar's argument rests on the evolution of "language" and its relationship to the size of the brain's neocortex. While spoken language underlies humans' obsession with gossip, Dunbar claims that grooming serves a similar function in some animals, especially our primate cousins, and that there is a close functional relationship between grooming and language. Available data seem to support this exciting and previously unexplored relationship.

Field observations show that individuals in many groups of primates stay busy grooming one another. Long sessions of grooming, occupying up to 10 percent of the time, have been recorded in some of the more social primates. When primates groom, they also chatter with one another. Touching and "talking" are two important components of the social systems of various primates.

The basic elements of Dunbar's argument are as follows: (1) In nonhuman primates, there is a close relationship between social group size and the size of the neocortex such that group size seems to be limited by the size of the neocortex; the larger the neocortex, the larger the average group size of a given species; (2) In humans, the size of social networks is limited to about 150 individuals; (3) The time devoted to grooming by primates is correlated with group size, and grooming plays a major role in the formation and maintenance of social bonds; (4) Language evolved in humans to replace social grooming because the amount of grooming needed to form and maintain social bonds was too great given other demands on individuals' time.

According to Dunbar, language evolved because it was more efficient for sharing information than sitting down with a friend and picking fleas out of his fur or playing with her hair. Furthermore, language allows for communicating with more than one individual at a time over a longer distance. Language also allows us to keep track of what other group members are doing and to advertise ourselves. It is also known that grooming is related to the release of opiates, indicating that grooming has a pleasurable component. Gossip can also be pleasurable.

So, do some nonhuman primates really gossip? Even if they do not gossip in the same ways that we gossip, it is entirely possible that grooming and chattering are functional equivalents and evolutionary precursors of human gossip in that each allows for the exchange of social information that is important in forming and maintaining social bonds. It is thought that grooming serves to build trust and individual knowledge about other group members, much the same as gossip. Perhaps, as Dunbar argues, language evolved as a form of "vocal grooming" to allow individuals to say something about themselves, to "groom" a large number of individuals simultaneously, and to bond to large groups. Indeed, Dunbar claims language actually evolved to allow humans to gossip, and much evidence supports his novel hypothesis. Who would have thought there would be close relationships among grooming, gossip, and big brains? Dunbar's research illustrates where broad and ambitious thinking can lead.

Do animals know what others know?

Animals spend a lot of time sleeping or just being lazy. In many cases 90 percent of their waking hours are consumed by just hanging out and watching others play, find food, eat, and groom. But do they learn anything about what other individuals know by being voyeurs? Researchers who study animal cognition, how animals think, want to learn if animals know about others' knowledge. It is not surprising that in some species individuals know a great deal about what others know. Knowing what others know is a perk of group living.

Recently Brian Hare, at Harvard University, along with Josep Call and Michael Tomasello, asked the question "Do chimpanzees know what other chimpanzees know?" Because chimpanzees rely heavily on vision to acquire information, the researchers wanted to learn if chimpanzees show an understanding of what others can and cannot see. Anecdotes suggest that chimpanzees and other animals are well aware of what others can see. Jane

Goodall observed a chimpanzee refrain from retrieving or even looking at fruit when other chimpanzees were present, only to retrieve it after others left. One of my former graduate students, Susan Townsend, discovered that wolves refrain from caching or retrieving food when other wolves are present. Chimpanzees will also hide parts of their body—for example, a facial expression called the "fear grimace"—so that others will not see they are afraid.

Scientists want more than fascinating stories, so Hare and his colleagues performed a set of clever experiments to learn if seeing leads to knowing. Chimpanzees can follow the gaze of another chimpanzee, so they potentially can learn something about what others know by watching the direction of their gaze. Hare and his colleagues set up a situation in which a dominant and a subordinate chimpanzee competed for food. Wild chimpanzees normally compete for food, so this is a natural situation; they did not have to be trained in an unnatural context. In some instances dominant chimpanzees did not see food being hidden. If they did, the food was moved elsewhere when they were not looking. Subordinate chimpanzees always saw the food being hidden or moved and could see what their dominant friends saw.

Hare and his colleagues discovered that subordinate chimpanzees were aware of what dominant animals did or did not see. Subordinates retrieved food that dominant chimpanzees had not seen hidden or moved. Hare and his colleagues also found that not only could subordinate chimpanzees keep track of what other individuals knew, but they could also keep track of *who* had seen what. When a dominant chimpanzee who had witnessed the hiding or moving of food was replaced with another chimpanzee who had not, subordinate chimpanzees knew that the naive chimpanzee did not know where the food was, and they retrieved it.

These and other experiments show that chimpanzees know what other group members have and have not seen, what they do and do not know, and that they use this information to make future decisions. Chimpanzees can take the perspective of other individuals and know that "others can see things that I cannot see and vice versa."

Many people might throw up their hands and say, "So what?" Is it not obvious that chimpanzees and other animals must know what others know so that they don't have to waste time and discover everything on their own? Yes, but what is exciting is that these "naturalistic" ecologically relevant studies support stories about wild chimpanzees.

Natural history has a very important place in studies of animal behavior. Similar studies on other species are needed, for it is unlikely that only chimpanzees are so smart. In many cases animals are as smart as our methods of study allow them to be. We just need to be clever enough to tap into how they do things in their worlds, not ours.

Self-medication in primates

Another intriguing activity is self-medication, or zoopharmacognosy, in which animals choose to eat plants that can help them control parasites and give relief from upset stomachs. Plant-secondary compounds and bark that are poor in nutrients and otherwise nonnutritional are ingested to provide such relief.

Michael Huffman, a professor at the Primate Research Institute at Kyoto University in Japan, has studied self-medication in chimpanzees in different East African populations. He discovered that some chimpanzees eat a plant that the local people know has medicinal effects. Once, a female named Chausiku fell ill. When others fed, she slept. However, at a later time when she was traveling with her troop, she stopped and intentionally peeled the bark off a mjonso tree and chewed on the pith. She then spit out the fibrous material and swallowed the juice. The bark of the mjonso tree is very bitter, and this was the first time that Huffman had seen a chimpanzee eat this plant. Huffman's local collaborator, Mohammed S. Kalunde, a national park game officer and herbal healer told him that it had medicinal qualities. Kalunde's people, the WaTongwe, use the plant to treat various gastrointestinal disorders including malaria, parasitic infections, and upset stomachs. The plant, in fact, is used widely across Africa by millions of people to treat many of the same symptoms displayed by Chausiku (and other chimpanzees) when she used the plant.

The fact that Chausiku was ill and was chewing on the bark of the mjonso tree was intriguing to Huffman, and as many keen scientists do, he put two and two together and got four. Indeed, Chausiku was self-medicating, practicing a form of animal medicine, and using the bark to help herself heal. The next day Chausiku was back to normal, eating ginger, figs, and grass.

Bonobos and gorillas are also practiced pharmacists. The fact that the same medicinal plant is chosen by nonhuman and human primates to cure similar illnesses might provide evidence for the evolution of what Huffman calls "medicinal behavior" in early hominids.

Another very interesting discovery entailed a comparison of different populations of chimpanzees. Apes of the same species who live in neighboring troops or in other populations tend to use many of the same or related species of plants. Different ape species also use many of the same or related species of plants. These observations suggest that all apes use some common criteria when they choose plants for self-medication. It is not known exactly what criteria are used for plant selection, but it is possible that the apes come to associate medicinal plants' rough hairy surfaces or odors with ingestion and feeling better.

Huffman notes that one of the most challenging questions facing future studies of zoopharmacognosy deals with how individuals acquire the habit. Not only do individuals have to chose the correct plant, but they also have to know which parts of the plant need to be ingested and how to obtain them. There are a number of possibilities, all of which might contribute to the acquisition of skillful self-medication.

First, choosing the correct plant and associated parts might be innate in that there is an inborn predisposition to select the right plant for a given illness. While this seems unlikely with such complex behavior patterns as plant selection, there would be a premium on doing it correctly the first time so that an illness does not progress to the point of being seriously debilitating or fatal. It might also be that naive individuals have the empathic ability to choose what they see others eat when they are sick. Huffman suggests that youngsters might learn what foods can help them feel better by watching what their mothers eat when they are ill. Indeed, infants have been observed to imitate their mothers immediately after they have fed on a particular medicinal plant. It is not only a matter of what she ate but how she ate it. It also might be that naive apes try different foods when they are ill, and when they feel better they associate their improved health with a particular food. Studies of taste-aversions have shown that many animals, even white rats, are able to associate the taste of a specific food with how their stomachs feel, so it is not asking too much of apes or many other animals to make such associations. Human infants regularly make these associations in the absence of knowing that they are doing so.

Huffman also has put forth his "Velcro theory" of leaf-swallowing. Jane Goodall observed chimpanzees swallow whole leaves early in her work at Gombe Stream. Huffman later noted that the leaves that are chosen have a bristly underside, and he has suggested that as the leaves pass through the intestine they catch worms and carry them along until they are excreted. Huffman has in fact discovered live worms in chimpanzee stools in the bristles of leaves that passed through the intestine.

As with many other interesting behavior patterns, there is an air of mystery surrounding zoopharmacognosy. How do apes and other animals learn what to eat when they are sick, and how do they come to associate a specific plant with a specific illness? What is the role of cultural tradition in the development and maintenance of plant choice? These questions are very difficult to study in the field because self-medication occurs rarely and unpredictably, it is very difficult to follow sick individuals over a period of time, and experimental manipulations are difficult to carry out. However, given that numerous apes and other animals are kept in zoos, more controlled experiments can be performed to learn about this intriguing but little-understood behavior. While these studies do not mean that zoos *have* to exist, clever experiments could be developed that enrich the captive environments for these individuals. Furthermore, the results could potentially be used to help the animals themselves. Huffman and his colleagues plan to continue to study zoopharmacognosy, for the mysteries surrounding it are interesting not only in and of themselves but also because we will likely learn something about our own ancestors.

Watching birds watch out for one another

Head up, head down, a shuffle to the right, a glance over a shoulder. What in the world are the birds doing, and why?

For more than ten years my students and I studied the behavior of western evening grosbeaks in the foothills of Boulder. One important grosbeak activity centers on avoiding becoming someone else's meal. Thus grosbeaks spend a good deal of time keeping their heads up and scanning for predators. They are vigilant. It is easy to know where they are looking because their eyes do not swivel. Their eyes (like those of most birds) move only when they move their heads. By recording the position of their heads we were able to judge whether they were looking around (their heads were up) or whether they were feeding (their heads were down) and were not seeing much of what others were doing.

Scanning helps individual birds learn about the presence of potential predators and also how many other birds are nearby and what they are doing. If other birds are scanning, an individual may decide to feed, but if others are feeding the individual might decide to scan. Individuals change their behavior based on what others are doing.

Many researchers are interested in how individual patterns of scanning vary in groups of different sizes. Generally, in many species of birds, as group size increases from two or three to eight or ten individuals, the likelihood that a predator will be detected increases also. As group size increases, the amount of time that an individual scans decreases. More eyes can be used for surveillance. Individuals also seem to coordinate their behavior, trading off feeding and scanning as if they are dancing with one another.

Group shape also influences scanning. By analyzing video frames in great detail, we found that grosbeaks scan more and change their head and body positions more often to see other flock members when they are in a line and it is difficult to see one another than when they are in a circle and other flock members can be easily seen.

Grosbeaks organized in a line also show less coordination in trading off scanning and feeding. Birds in a line are trying harder to see what others are doing, most likely to decide what *they* should be doing. "If others are

A male western evening grosbeak takes a break from feeding and scans his surroundings, gathering information on what other group members are doing—are they also scanning, or are they feeding?—and whether there are potential predators lurking.

scanning, then I can feed, and if others are feeding, then I can scan." When in a line, other birds interfere with an individual's ability to see who is there and what they are doing. Grosbeaks need to see others because they change their behavior based on what they see. Thus they move around a lot.

Grosbeaks (and other animals, including the deer who regularly visit my home) are busy and nosy because they spend a lot of time scanning for predators and trying to see how many other group members are present and what they are doing. They behave flexibly depending on how many other birds are around and how they are organized in space. Flexible, adaptable response to changing and unpredictable environmental conditions is the major criterion used to indicate the presence of active, thinking minds.

Some people might fear that we are attributing too much thinking or brain power to birds. But it is important to note that researchers have found that many birds form mental representations of a moving stimulus as it disappears and later reappears, and by keeping track they know that it is the same image. This ability is important for keeping track of prey that appear, disappear, and then reappear in a different location. Some birds, such

A circle of western evening grosbeaks in which each bird is able to see all other individuals in the group.

as African gray parrots, also develop concepts about the shape of objects and make discriminations based on number. All of these skills are related to scanning behavior. Interestingly, relationships between scanning and group size usually become unpredictable in flocks of greater than ten birds. Whether or not birds are estimating the number of individuals present (referred to as "subitizing") or actually counting the number of other individuals in a flock awaits further research. One thing is clear, calling someone a birdbrain is not necessarily a criticism of the person's cognitive capacities.

Deciding where, what, and with whom to eat

When I am hungry I often ask myself, "Where, what, and with whom shall I eat?" I also want to know if there is there anyplace or anyone I want to avoid. Many other animals seem to ask these questions too.

A common bird that lives around my house is the Steller's jay. Its deep blue color, large, regal head crest, and annoying squawk make it difficult to ignore. Recently Colin Allen (at Texas A&M University), Ann Wolfe (at the University of Wisconsin), and I studied feeding patterns in these birds to learn about the mealtime decisions they make and why they make them. We discovered that many different interacting factors influence feeding. There were few simple answers to the questions of where, on what, and with whom to dine.

In our studies birds were fed sunflower seeds, safflower seeds, or a mix of the two and were filmed on two feeding platforms. Jays were most attracted to sunflower seeds. Jays also paid a lot of attention to other jays. When another jay was on the same platform, but not when it was at some distance on the other platform, a jay pecked at a lower rate than when it was alone, because it was busy watching the other bird.

The simple relationships we first discovered could have lulled us into thinking that feeding by jays was pretty straightforward and that we were very clever for figuring it all out. We could predict what jays would do if we knew seed type and whether or not other jays were around.

But the story is not all that clean. Additional studies showed that feeding patterns are simultaneously affected by many variables, including seed type, the presence of other jays and even fox squirrels, the relative dominance of the birds, and the distance from the feeding platform to the protective cover of nearby trees. Jays were making complex decisions after evaluating the total situation, not just one variable at a time.

We found that the presence of other animals could override a preference for sunflower seeds. They were not all that appealing when other jays, especially more dominant individuals, were present on a feeder. We could not have known this if we had only considered seed type but ignored possible interactions between seed type and the presence of other jays, especially dominant birds.

Jays also selected the unoccupied feeder over one occupied by another jay or a squirrel, with squirrels being avoided more than other jays. Jays also usually chose the feeder further from cover, possibly because it was more accessible; there were more arrival and departure routes. The more open feeder also might have allowed jays more easily to watch other animals. Jays did not suffer much predation, so being more exposed did not seem to have any negative effects.

Jays, like many other animals, make complex choices by simultaneously evaluating various stimuli. Studies and models that consider how various stimuli are integrated are more realistic than those that only consider a single stimulus. You can repeat these types of experiments with jays and other animals to learn more about their worlds, how they use all sorts of input to make choices right in front of your eyes.

Dreaming rats

Most of us have watched dogs run in their sleep. Often their running in place is accompanied by growling or barking. We imagine that they are dreaming of chasing a cat or some other animal. But can we really believe that rats dream?

Kenway Louie and Matt Wilson, working at the Massachusetts Institute of Technology, have discovered that when rats sleep they appear to dream about a maze in which they ran the previous day. Rats were trained to run along a circular trace for food rewards. Louie and Wilson recorded electrical activity (taking an EEG, or electroencephalogram) from cells in the rats' hippocampus, an area of the brain important in forming and storing memories. They found that there is a distinctive pattern of firing when the rats are learning to navigate through a maze, and the same firing pattern is found when the rats are in rapid eye movement (REM) sleep, when dreaming occurs. In fact, the correlation in the EEGs when the rats were in the maze and when they were dreaming was so close that Louie and Wilson could figure out where in the maze dreaming a rat was and whether the rat was dreaming of standing still or running.

What it all comes down to is that dreaming is not a uniquely human phenomenon. As with tool use, language, culture, and self-awareness, to name but a few behavior patterns that were once thought to be uniquely human, dreaming can be added to the list of activities that are shared by many animals. This does not mean that dreaming in rats or other animals is the same as dreaming in humans, but it also is too early to know if and how animal dreaming differs from human dreaming. Antii Revonsuo has recently suggested, based on content analyses of human dreams, that human dreaming may have evolved to simulate threatening events, as a way to rehearse threat perception and to develop and maintain behavior patterns that are important in threat avoidance skills. He argues that dreaming is not a random by-product of REM sleep and that dreaming may serve this natural function. Perhaps this is also true for animals. Marek Spinka, Ruth Newberry, and I have argued that play behavior may be important for training youngsters for unexpected circumstances, some of which may be threatening. The relationship between dreaming and playing remains to be sorted out.

Drafting pays off: Lessons from geese

As a bicyclist, I know that drafting another rider, referred to as "sucking his wheel," saves energy. But what about other animals? Why do aquatic animals swim in a line? Why do geese and other birds fly in spectacular V formations?

Two hypotheses have been offered to explain V formations. One posits that it saves energy, and the other maintains that it facilitates communication. Thus birds who fly long distances can exchange information about the most suitable feeding grounds, details of which might help younger and inexperienced birds find food.

When birds fly in V formation, usually only a single bird flies in front with others following on either side of the leader. There is no evidence that leaders and followers regularly trade positions as they do in an echelon of bicyclists, nor that birds synchronize wing-flapping to maximize energy gains.

Do followers save energy when drafting, or is information more easily exchanged among birds flying in a V? Detailed research conducted by Professor John Speakman and a colleague in the zoology department at the University of Aberdeen in Scotland indicates that greylag geese, the largest native goose in Europe, do indeed show energy savings when flying between their roosting and feeding grounds. They filmed twenty-five formations, or skeins, involving 272 geese, from below so as to minimize distortion. By carefully analyzing the films, they discovered that the geese flew very close to one another (but kept enough distance to avoid collisions) and that the degree of overlap of their wing tips agreed well with an aerodynamic model that predicted the position at which energy savings would be maximal. Although only 17 percent of the geese flew in the most optimal position in the V, mean energy savings among all birds in different positions was about 27 percent, and there was an average savings of 4.5 percent to 9 percent in the total energy costs of the flight. Of great significance, 97 percent of all geese showed some energy savings from flying in formation.

It has also been suggested that energy savings are increased for larger birds flying in formation when compared to smaller birds, because larger birds have higher energy demands when flying. Indeed, smaller pink-footed geese do not show energy savings as great as their larger greylag kin. Pink-footed geese show mean energy savings among all birds in different positions of about 16 percent and save only 2.7 percent to 5.5 percent of total flight costs. They could have saved considerably more energy by flying in different positions, if saving energy was the reason they flew in formation. Nonetheless, it may be that even minimal savings are important for certain individuals, especially weaker or younger geese.

Now, what about humans? Dr. Andy Pruitt, my close friend who is Director of the Boulder Center for Sports Medicine and whose wheel I and others have mercifully sucked on countless occasions, tells me that depending on local conditions—wind, position in a line, and speed—energy savings for drafting cyclists and speed skaters can range as high as 30 percent, while those for runners can be as high as 10 percent. Energy savings increase with speed.

Social dominance

"Peck to the left, step to the right, rear up, dive in, growl, bite, fight." This jingle, written on one of my bike rides home from the university, describes what many animals do when they establish dominance relationships.

Many animal groups display dominance hierarchies in which individuals socially control others. Hierarchies are often called "pecking orders" because early studies of dominance were conducted on chickens, who peck at one another during aggressive encounters.

Dominance relationships can be formed by fighting. But fighting is risky, so threatening often works. Growls, bluffs, direct stares, and fluffed-up hair or feathers can achieve the same results—submission, withdrawal, or avoidance—as a down-home brawl.

Individuals also might use others' size to decide if it is best to fight, to threaten, or to leave town. In many species, larger males usually dominate smaller males and females. However, in house finches and spotted hyenas, females generally dominate males, regardless of relative size. In chaffinches, during breeding, smaller females dominate males. Indeed, in many species, females control males during breeding.

Some researchers feel that male bias—there have been many more male than female researchers—leads to conclusions that males dominate females. Some studies show that where males see dominance, females see equality. In chapter 2 I noted that female researchers frequently see cooperation when males see competition in the same social encounters.

What is beneficial about being dominant? While there are variations within and between species, dominant animals, when compared to subordinates, generally are freer to move about, have priority of access to food, occupy more protected parts of a group, attract higher-quality mates, suppress others' reproductive activities, and control the attention and behavior of group members.

But being dominant may also be costly. Dominant animals often suffer from stress associated with the anxiety of being overthrown or from being ostracized from their group. Or, when dominant males chase off competing males, others can sneak in and copulate with their mates.

Now, the big question. Has dominance evolved because dominant individuals mate more and also produce more young over their lifetime? Certainly, many humans believe that the answer is a resounding yes, at least concerning mating. However, in many species, researchers have not found high positive correlations among dominance status, sex life, and lifetime reproductive success.

Establishing causal links among dominance status, sex life, and lifetime reproductive success is very difficult. Because age and size often are highly correlated with rank, it is essential to separate out their impact on reproductive success. In baboons, high female rank is associated with elevated levels of testosterone that can inhibit pregnancy. DNA fingerprinting has shown that multiple paternity can occur in animals including dogs, pigs, and swallows. It is also very hard to follow identified individuals in nature for a long time.

For dominance to evolve, its costs—energy, risks, or stress—need to be less than the reproductive benefits to an individual or relatives. In male and female African wild dogs, paper wasps, red deer, and savanna baboons, and in male copperheads, elephant seals, and sage grouse, dominance is highly correlated with mating success. In spotted hyenas, high-ranking females, compared to low-ranking females, give birth at an earlier age, and more of their offspring survive to adulthood. In chimpanzees, infants of high-ranking mothers are more likely to survive, and daughters of high-ranking mothers mature more rapidly than daughters of lower-ranking mothers.

There is still much to learn about dominance in animals. Thus it seems unwise to use animal examples to rationalize such activities as brawling or strutting for reproductive advantage by humans. In many species males brawl and strut, but females choose.

Sexual swellings in female baboons

It is not always the case that females do the choosing and males do the competing and strutting. In olive baboons, females display what are called "sexual swellings" around the perineal skin when they are ovulating. It turns out that these swellings—bigger bottoms—advertise the quality of a female to males. Leah Domb and Mark Pagel discovered that females with larger swellings attain sexual maturity earlier in life than females with smaller swellings, produce more offspring and more surviving youngsters per year, and have a higher proportion of offspring survive. Basically, females with larger swellings are more reproductively fit in the Darwinian sense of the word in that they have higher reproductive value over the course of their lifetime. Male olive baboons have to be selective in the mates they choose, and they are more interested in, fight more to mate with, and spend more time grooming females with larger swellings. Males benefit from being choosy because it is more likely that their offspring will survive if a female with a large swelling gives birth to them. Another very interesting discovery was that the size of the swelling is independent of a female's social rank and age.

The ways we choose the mates we do

What does it mean to say someone is "attractive" or "beautiful"? How much control do we have over our assessments of beauty?

These questions are receiving a lot of attention from biologists and psychologists. Much information is reviewed in Nancy Etcoff's *Survival of the Prettiest: The Science of Beauty* and Victor Johnson's *Why We Feel: The Science of Emotions.*

Are you bilaterally symmetrical—are the right sides of your face and body mirror images of the left? According to some researchers, you will be more attractive if you are well balanced. This is so for other animals. Many female fish prefer males who have symmetrical markings. Female black grouse prefer males with the most symmetrical feet, and female house sparrows prefer males with larger, more symmetrical bibs.

Now, what about humans? Females generally prefer the scent of men with symmetrical bodies over that of asymmetrical males, and when females are most fertile, they prefer more symmetrical men. (The menstrual cycle can alter face preferences.) Females also report having more orgasms with symmetrical men, which means they are more likely to conceive with well-balanced lovers. Symmetrical men report having sex at a younger age, earlier in courtships, and with more partners. Symmetrical females also report having more partners.

It is speculated that being symmetrical means having "good genes" and being better able to resist stress. If an individual's health and competitive abilities (strength, endurance) are related to his or her degree of symmetry (as they are in horses, birds, and insects), then symmetry might be the litmus test for one's ability to handle stress. Asymmetrical individuals in many species have lower survival and growth rates and produce fewer kids.

Here is more interesting information. Across numerous diverse cultures (but not all) and also historically, men find women with waist-to-hip ratios of about 0.7 (the waist is seven-tenths the size of the hips) to be most beautiful. Full lips and a small jaw also figure into the equation, but breast size and weight are less important. The hormones that influence waist-to-hip ratio also affect facial features. Deviations from 0.7 are associated with marked reductions in fertility, and small flat waists are associated with proper hormone functioning and good health.

Many of our choices are mostly unconscious decisions. People often cannot explain the reasons for their preferences. Perhaps because many scientists prefer explanations based on hard-and-fast rules, they are prone to accept theories that make it look like our choices are mechanical. The heart and soul of our decisions, the soft-and-slow reasons underlying our preferences, are pushed aside for explanations that jibe with blind, heartless evolutionary theory.

More information is needed concerning how we perceive beauty. Etcoff correctly notes that "we have to understand beauty, or we will always be enslaved by it." Do we really have little control over our assessments of beauty and our genes' futures? Do assessments of beauty reside in our evolutionary past, in individuals' hearts or noses, in body language, in our genes, in features beneath the skin, in individuals' idiosyncratic notions of what's beautiful, or in the agendas of big businesses that cultivate it? The best guess is that assessments of beauty stem from some unknown combination of all these factors and then some.

Females prefer attached men

I am sure that on many occasions most of us have looked at couples and thought something like "Wow, what does she see in him?" or "What does he see in her?" Sure, this borders on being petty, but everyone is narrow-minded at least once! Recent research by Lee Dugatkin, a biologist at the University of Louisville in Kentucky and author of the book *The Imitation Factor: Evolution beyond the Gene*, has discovered a phenomenon called "mate-copying" in guppies that might also explain patterns of attractiveness in humans.

"Mate-copying" refers to the situation in which an individual is more likely to have an interest in a member of the opposite sex if he or she is already paired up. Dugatkin found that a female guppy would actually change her mind about a male if other females preferred him.

Female guppies usually prefer brightly colored orange males over drabber fish. Bright males are more brazen and will approach interlopers and predators. But Dugatkin did not stop here, and his persistence, curiosity, and cleverness paid off. Dugatkin conducted an experiment in which a bright male and a drab male were kept in small tanks on either side of a larger tank. He then placed a female (let's call her Mary) into the middle tank; as predicted, Mary courted Mr. Bright. But when another female (Sally) was put in the middle tank and was forced to swim near the drab male, and Mary, who had preferred the bright male, was held in a cylinder so that she could see Sally court the drab male, Mary courted Mr. Drab rather than Mr. Bright when she was released into the tank. Mary changed her preference from Mr. Bright to Mr. Drab because Sally preferred Mr. Drab.

Mate-copying has also been observed in Japanese quail and other fish including mollies and swordtails. Why individuals engage in mate-copying remains a mystery, but it might be that it saves time and energy that could be put into other activities such as feeding, resting, or avoiding predators. Dugatkin found that young guppies are more likely to copy older guppies than older guppies are to copy younger guppies, so it might be that copycats (or copy-guppies) are taking advantage of what others have learned in the past.

What about humans? Here is where it gets interesting. Dugatkin and his Louisville colleague, psychologist Michael Cunningham, discovered mate-copying in humans. In their study, men and women all tended to prefer attractive members of the opposite sex, but women were more likely than men to pay attention to preferences of members of the same sex. This observation might help to provide an answer to the questions "Wow, what does she see in him? and "What does he see in her?" What is seen is the preferences of others.

Mate-copying in humans remains a mystery. However, female humans and other animals tend to trust other women more than men trust other men in assessing whether members of the opposite sex have "something good" going for them. Because females in some species have more to lose than do males by choosing the wrong mate, perhaps "mate-copying" is another example of female choice, in this case females preferring males who are already taken.

The distribution of mate-copying among different species awaits further study. So do more detailed explanations. But it is possible that the simple explanations really work the best. If one female finds another male attractive, then other females will also find that male attractive.

Let me add one more observation. In many species female-female aggressive encounters are more explosive than male-male aggressive interactions. This is so in golden jackals and coyotes, and Christine Drea at Duke University and Patty Gowaty at the University of Georgia have told me that it also is the case in spotted hyenas and eastern bluebirds, respectively. In hyenas female-female aggression can be deadly.

Perhaps it is because "a good man is hard to find"—a guy who can ward off interlopers and competitors and also provide care for his mate and their young—that females will compete for that special guy. In the end there might be a good reason why females trust one another. Why not let others help you make difficult decisions when the consequences of a mistake—picking an irresponsible guy who can't or won't provide care and protection—are high?

Why help to rear others' kids? Kin selection and reciprocity

In many animals, adults other than parents provide care to youngsters. Even Charles Darwin, perhaps the most influential biologist of all time, was perplexed by this phenomenon. Why should individuals help rear others' offspring rather than reproduce on their own? How might such unselfish behavior—altruism—evolve?

Males and females who provide care to others' young are called "helpers." Helping behavior is also called "alloparental" behavior. Helpers may incubate eggs, provide food to young or parents, build and defend nests and dens, defend territories, or just hang out as sentries and baby-sit. Their presence may keep predators and other intruders at bay.

There are several theories about how helping evolved. One, "kin selection," is the evolutionary process by which relatives come to favor kin over nonkin. Individuals who favor kin typically have higher reproductive success than those who do not, and the preference for kin is selected evolutionarily over time. Individuals are not necessarily making conscious decisions, so the word "selection" can be misleading.

In birds and mammals, many cases of helping are explained by kin selection. Relatives receive more help than nonrelatives because they share a certain percentage of genes with the helper. Thus some of a helper's genes are carried indirectly into subsequent generations via relatives, and the helper's reproductive fitness is also increased. A helper may also inherit a territory on which to breed.

Another theory of how helping may evolve concerns the notion of reciprocity: I will help you now because I hope you will help me in the future. I am willing to wait for the delayed benefits. I am betting that you will not cheat and that you will remember what I did for you in the past. Reciprocity can explain why nonrelatives might help one another. It does not matter if it is kin or nonkin who pays you back.

With a team of dedicated students, I studied coyotes in the Grand Teton National Park, outside of Jackson, Wyoming, for seven years. Coyote helpers mainly provided care to offspring by defending territories or den sites from intruders or by baby-sitting when parents were off hunting. One male, Bernie, remained with his pack for three years and helped to rear younger siblings. Bernie could not breed when his father was present, so helping was the next best activity in lieu of reproducing on his own. When his father left the pack, Bernie inherited the pack's territory and mated with a female who joined the pack after his mother left. Bernie, his mate, and his offspring received help from coyotes to whom Bernie had previously provided care. Because Bernie had previously provided help to relatives and to individuals from whom he and his pups received care later on, his helping can be explained by both kin selection and reciprocity.

It is known that helpers make a difference in the survival of offspring through at least infancy. In many species a helper's presence increases the survival of young and the reproductive success of parents, especially when there is enough food to go around. In coyotes, on average, two helpers are responsible for the survival of one additional pup. In African wild dogs, wolves, Galapagos hawks, pied kingfishers, Florida scrub jays, and acorn woodpeckers, helpers' presence also increases the survival of young. However, in wolves and red foxes, when there is not enough food for everyone, helpers' presence can actually lead to a decrease in the survival of youngsters. What we really do not know on any broad taxonomic scale is whether the contributions of helpers to the increased survival of the young to whom they provide care is a statistically significant effect. Only long-term and difficult field studies will help us here, and they are, for the most part, lacking. Until these data are available, helping looks to be a "biologically significant" activity whose actual long-term contributions remain unknown. Nonetheless, the trends that have been found in a wide variety of species suggest that helping has benefits that are yet to be discovered.

Had Darwin known about kin selection and reciprocity, helping might have been less of a mystery. Helpers are not always being unselfish altruists, because their own reproductive fitness may also increase.

It is the variability in helping behavior that drives new research. No single explanation applies in all situations. Researchers want to know why some individuals become helpers and some do not, why helping is observed in some species but not others, and what ecological conditions (mainly food) favor helping. This information might explain the existence of indulgent aunts and uncles in humans.

Do dogs know when their human companions are returning home?

Rupert Sheldrake and his colleagues have repeatedly claimed that some dogs know when their human companions (aka owners) begin returning home, even if they return randomly. They have performed a number of different experiments, but most people remain very skeptical of their findings because Sheldrake posits that there is a telepathic connection between the dogs and the humans. Recently, Sheldrake and Pamela Smart reported the results of ten videotaped controlled trials in which Kane, a male Rhodesian ridgeback, was observed to spend an average of 26 percent of the time at the window of his house while his owner was returning home, whereas he spent only 1 percent of the time at the window when she was not. The window area of the house was continuously filmed while Kane's owner traveled to places more than eight kilometers away and returned home at random times after being away from 130 to 330 minutes. Furthermore, the videotapes of Kane's behavior were scored blindly by a third party. Sheldrake and Smart ruled out the possibility that Kane heard or smelled his owner because Kane began looking out the window when she was more than seven kilometers away.

Do some dogs have telepathic powers? Sheldrake's study and reports by many dog owners suggest that they do. We do not know much about whether or not wild animals anticipate the arrival of group members even if they return after being away for random amounts of time. I mention Sheldrake and Smart's study because it appeared in a peer-reviewed professional journal, the standard by which researchers are judged. Future research is needed to determine whether explanations that invoke telepathy are the most plausible accounts for return-anticipating dogs.

four

ANIMAL MINDS AND
WHAT'S IN THEM

Thinking bees

Ever since Karl von Frisch's classic studies of the ways in which bees tell other bees where they can find food using the waggle and round dances (for which he shared the 1973 Nobel Prize), people have debated whether or not bees (or other animals) have "language." If by "language" we mean human language, the answer is no. Nonetheless, bees and other animals perform very complex behavior patterns in the absence of human-like language. Now there is evidence that honeybees can reason, that they know the concepts of "same" and "different," abilities that were previously reserved for vertebrates, mainly primates. Martin Giurfa and his colleagues took advantage of the fact that honeybee workers can search for and find honey, return to their nest, tell others where the best source of honey is, and then return to this source. Put another way, bees can "interpolate visual information, exhibit associative recall, categorize visual information and learn contextual information," abilities that are taken to be evidence of thinking. Giurfa and his colleagues trained honeybees to recognize colors and grating patterns in a Y maze. When the bees entered the maze they saw either blue or yellow, and when they arrived at the Y junction they saw that one direction was labeled blue and the other yellow. Bees were able to learn that the reward of a sugar solution was found in the arm of the maze that was the same color as the entrance. They could learn the same task with similar and different grating patterns and also when lemon and mango odors were used instead of colors. Because the bees could tell sameness from difference, the researchers concluded that they could think. The same conclusion has been drawn for other animals who perform similarly. Thinking is no longer a feature of only some vertebrates.

In the previous chapter I discussed many different aspects of animal behavior to demonstrate the diversity and flexibility with which animals interact in their social and nonsocial environs. Now I turn to a general discussion of animal minds. There is much interest in this area of research, not only because people want to know just how smart other animals are, but also because discovering information about animal minds and what is in them is closely related to discussions of animal emotions, the evolution of social morality, the ethics of animal use, the effects of human intrusions on animal behavior, and spirituality, topics to which I will devote considerable attention.

I will also consider consciousness and self-consciousness (also referred to as self-awareness). These are challenging topics that raise hackles among some of my colleagues, not only because they are fraught with difficult issues about which very little is known, but also because researchers use these words inconsistently and, in my opinion, in too sweeping a manner. I will discuss consciousness and self-consciousness from comparative evolutionary, social, and ecological perspectives, asking why each may have evolved—what are they good for? This approach removes much of the mystery that some find unsettling. I will also question whether some species have a "need to know" who they are in order to function as normal members of their species. Behavioral flexibility often is given as one of the main reasons why animals might need to consciously process information. There is much research being conducted on the neural bases of behavioral flexibility and consciousness, and this rich and growing field could be a book unto itself. Researchers are discovering intriguing relationships between, for example, such variables as forebrain size and feeding innovations and behavioral flexibility in birds and between the size of the brain relative to the size of the body and behavioral flexibility and sociality in mammals. There is some truth to the claim that relative brain size and behavioral flexibility are positively correlated, but much more work needs to be conducted in this fascinating area.

Comparisons between animals and humans

Do animals think, make plans, or have beliefs about others' behavior and intentions? Do they try to deceive one another? Do they have likes and dislikes? Do they feel joy and sorrow, pleasure and pain? Do they know who they are? Do they have a "theory of mind"—do they know what others know or feel? Available information suggests that some chimpanzees and perhaps some other great apes and monkeys have a sense of self, dogs and many other animals make plans for the future, and many animals experience joy, sadness, and sorrow and feel pleasure and pain.

In addition to learning about the cognitive abilities of animals, some researchers are interested in making comparisons between the cognitive abilities or cognitive "levels" of animals and humans. For example, it is often suggested that the behavior of chimpanzees and the behavior of human children of about two and a half years of age are comparable with respect to language skills. I am not sure that it is very useful to claim that a chimpanzee can reach the "intellectual" level of a two-and-a-half-year-old human infant. Neither will we learn much new by continuing to rear chimpanzees as if they are human. These so-called cross-fostering studies tell us little if anything about the behavior of normal chimpanzees and raise numerous ethical questions. Each organism does what it needs to do in its own world, and surely a young human (or most humans at any age) could not survive in the world of a chimpanzee.

Along these lines, one of my colleagues in the sociology department at the University of Colorado, Leslie Irvine, told me an interesting story of her dog, Skipper. There are some lessons to be learned here concerning cross-species comparisons. Leslie and Skipper were playing near their house by a drainage canal that was running pretty swiftly. Leslie was throwing a stick for Skipper to retrieve, and the stick went into the running water. Skipper does not like to swim, but he will wade into shallow water. Skipper looked at the moving water and decided to run downstream to get ahead of the stick. He then waited for the stick and waded in and retrieved it. Leslie notes that Skipper understood some principles of basic causality and action that might go like this: "The water is moving in that direction. If I move in that direction, too, I can get the stick." Leslie and I (and I expect many readers) have seen young children in similar situations say something like "Uh-oh, all gone," not realizing that the stick was merely moving away and that it could be retrieved downstream. While there may be other explanations for Skipper's behavior, I am not sure what I would discover if I were told that children of a certain age usually develop the same ability that Skipper displayed and that Skipper was as smart as a child of that age, but no smarter.

Cognitive ethology

Many researchers who study animal minds, especially in field situations, refer to themselves as "cognitive ethologists." Cognitive ethology is the comparative, evolutionary, and ecological study of animal minds and mental experiences including how they think, what they think about, their beliefs, how information is processed, whether or not they are consciousness, and the nature of their emotions. "Comparative" means that researchers compare and contrast closely related (for example, wolves, coyotes, and domestic dogs) and more distantly related species (wolves and lions or tigers, all carnivores); "evolutionary" means researchers want to know why and how a given behavior evolved over time; and "ecological" means they want to learn how variations in the food, the social environment, or the habitat influence the behavior patterns they are studying.

Cognitive ethology traces its beginnings to Charles Darwin. A natural historian at heart, Darwin emphasized the importance of evolutionary mental continuity among animals and learning about the worlds of the animals themselves when he wrote about animal minds. The notion of continuity is very important. Biologists freely talk about the evolution of kidneys, stomachs, and hearts from simpler to more complex systems, and we should feel equally comfortable talking about the evolution of brains, where mental processing takes place and minds reside. Of course, animal brains are organized differently from human brains, but this is not surprising and does not mean that animals' brains cannot house their minds and underlie such abilities as perception or self-awareness. It is narrow-minded to believe that we are the only species with minds or the only species that can think, make plans, and experience pain and pleasure.

Modern interest in cognitive ethology began with Donald Griffin's book *The Question of Animal Awareness: Evolutionary Continuity of Mental Experience*, first published in 1976. Griffin's major concern was to learn more about animal consciousness and how it varied in different organisms. He wanted to come to terms with the difficult question of what is it like to be a particular animal. Griffin stressed that it is the flexibility and versatility of behavior that provide strong evidence of animal minds and consciousness. He and others emphasized that animals, including humans, do not always think about what they are doing or going to do. However, when the environment changes and animals need to adjust their behavior to new situations, or when some fine-tuning is necessary, thinking and planning likely are used. Studying how individuals adapt to novel and rapidly changing situations is exciting and challenging and provides much information about what is happening between their ears.

Different views on cognitive ethology

In 1997 I published a paper with Colin Allen in which we identified three major groups of people with different views on cognitive ethology, namely, *slayers*, *skeptics*, and *proponents*. Although there often are blurred distinctions among these groups, by categorizing different views in this way we were able to clarify the reasons that are given for rejecting, remaining skeptical about, or accepting research in cognitive ethology.

Slayers: Slayers deny any possibility of success in cognitive ethology. We found that they sometimes conflate the *difficulty* of doing rigorous cognitive ethological investigations with the *impossibility* of doing so. Slayers also often ignore specific details of work by cognitive ethologists. They do not believe that cognitive ethological approaches can lead, and have led, to new and testable hypotheses. They often pick out the most difficult and least accessible phenomena to study (for example, consciousness) and then conclude that because we can gain little detailed knowledge about this subject, we cannot do better in other areas.

Slayers also deny the usefulness of cognitive hypotheses for directing empirical research. For example, the noted primatologist Lord Solly Zuckerman, in his review of Dorothy Cheney and Robert Seyfarth's book *How Monkeys See the World: Inside the Mind of Another Species*, exemplifies this dismissive attitude of slayers as follows: "Some of the issues they do raise sound profound as set out but, when pursued, turn out to have little intellectual or scientific significance." I wonder if Lord Zuckerman read the same book as others, even critics, who still see value in Cheney and Seyfarth's groundbreaking research.

Cecilia Heyes, who is a laboratory psychologist and also a slayer, agrees with Lord Zuckerman and has gone as far as to advise cognitive ethologists to turn to laboratory research if they want to understand animal cognition. Heyes once wrote: "It is perhaps at this moment that the cognitive ethologist decides to hang up his field glasses, become a cognitive psychologist, and have nothing further to do with talk about consciousness or intention."

Heyes denies that evidence gained by observing animals in natural settings is relevant to understanding animal minds. She and other critics generally simply assume that no evidence that could be collected from the field would provide convincing support for attributions of mental states. Like many of my colleagues, I believe that she and other slayers are wrong and that there is ample room for all types of research on animal minds. Indeed, if we do not know about the behavior of free-ranging animals, we will be unable to assess the validity of information that is collected in captive conditions.

Skeptics: Skeptics are often difficult to categorize. They tend to be a bit more open-minded than slayers, and there seems to be greater variation among skeptical views of cognitive ethology than among slayers' opinions. However, some skeptics recognize some past and present successes in cognitive ethology and remain cautiously optimistic about future successes.

Many skeptics appeal to the future of neuroscience and claim that when we know all there is to know about nervous systems, cognitive ethology will be superfluous. (Griffin also makes strong appeals to neuroscience, but he does not believe that increased knowledge in neurobiology will cause cognitive ethology to disappear.) Skeptics also do not favor anthropomorphic explanations.

Proponents: Proponents recognize the utility of cognitive ethological investigations. They claim that there are already many successes, and they see that cognitive ethological approaches have provided new and interesting data that also can stimulate further study. Proponents also accept the cautious use of anecdotes or anthropomorphism. If cognitive ethology is to die, it will be of natural causes and not as a result of hasty and often nasty slayings.

As proponents, Colin Allen and I argued that there are many reasons for studying cognitive ethology from comparative, evolutionary, and ecological perspectives. The assumption of animal minds also leads to testable hypotheses about, and more rigorous empirical analyses of, behavioral flexibility and behavioral adaptation. My own research on social play and antipredatory behavior in birds has benefited not only from collaborating with philosophers but also from taking a more cognitive approach. Studying animal cognition is not easy. It takes time and patience, and, as Sonja Yoerg noted, "It isn't a project I'd recommend to anyone without tenure."

What does "being smart" mean?

Are animals smart? Perhaps a better question is "Are *some* animals smart, and do they use their intelligence *some* of the time?" Nonhuman and human animals frequently act without thinking, or at least they act without knowing they have actually thought about what they are doing.

Let me offer a short and simple working definition of what it means to be smart. Being smart means showing versatile and adaptive behavior in novel and unpredictable situations and anticipating and planning for the future. Animals are said to be smart if they perform such tasks as counting objects, forming concepts in which differences or similarities are recognized, avoiding cagey predators, locating hidden food, making and using tools, or using complex forms of communication.

Animal communication

Studying communication is helpful for learning about intelligence and what might be happening in the minds of animals. Donald Griffin believes that studies of animal communication provide a "window" into the minds of animals. Con Slobodchikoff, at Northern Arizona University, has discovered that Gunnison's prairie dogs use distinct alarm calls to designate predators such as hawks, coyotes, and humans. These rodents clearly know who is who. Dorothy Cheney and Robert Seyfarth found that vervet monkeys announce differently such predators as pythons, leopards, and eagles, and group members respond appropriately to each call even when predators are unseen. Alex, an African gray parrot studied for years by Irene Pepperberg, understands the concepts of "same" and "different" and can answer questions about the number of objects present and their color, shape, and composition. Pepperberg discovered that when Alex is presented with plastic objects—three yellow, one purple, and one red—and one piece of green wood and asked, "What material green, Alex?" he answers, "Wood." And when asked, "How many yellow?" Alex says, "Three."

Kanzi, a bonobo who has been studied by Sue Savage-Rumbaugh, communicates with humans using a keyboard (lexigram) containing numerous symbols. Because he cannot talk when he wants something or is asked to respond to a question, Kanzi points to a symbol on the lexigram. Kanzi also comprehends spoken sentences such as "Take the vacuum outside" or "Open the pickle jar."

But what does this all mean? What do bonobos do in the wild? Is there any relationship between what Kanzi does in his pampered quarters and what wild bonobos do as they try to survive in their natural environs? Do they "talk" to one another in "Bonobo-ese"? In the wild, Kanzi's free-ranging relatives live in dense forests and track one another by leaving signs indicating where they are heading. When trails cross, leaders stamp down vegetation or place large leaves on the ground indicating travel direction. Trail notes are left only where trails split or cross and confusion can occur. When all members travel together, trail markings are not used, because all the bonobos can see one another.

Bonobos seem to be mind-reading. They appear to anticipate what other bonobos will do after seeing symbolic trail markers that are left for them to see. Even a human could follow the signs that bonobos leave behind: Savage-Rumbaugh was able to locate the hidden group using bonobos' trail signs. This is a nice example of how research on captive animals can complement studies on their wild relatives. The importance of studying animals in their own worlds has been stressed by Savage-Rumbaugh: "I believe it is time to change course. It is time to open our eyes, our ears, our minds, our hearts. It is time to look with a new and deeper vision, to listen with new and more sensitive ears. It is time to learn what animals are really saying to us and to each other."

Animals can also be deceptive. There are many examples of animal deception, especially about food. Deception includes concealing by silence or hiding an object, distracting others' attention by looking away from an object, or leading others away from an object. When rhesus monkeys find food, they will often withhold this information and not call others to the area. Susan Townsend, one of my graduate students, discovered that wolves often only cache and retrieve stored food when there are no other wolves around. This so-called audience effect is also observed in other species in which individuals change their behavior depending on who their audience happens to be. Chimpanzees have been observed to point in the wrong direction to humans, who tend to take all available food for themselves.

Deception also is observed in adult birds who are protecting their young. Carolyn Ristau discovered that female piping plovers feign a broken wing and hobble away from nests to distract a predator's attention. After the predator has been lured away, the mothers rush back to their chicks. I was once hiking in the Snowy Range in southeastern Wyoming when I saw a white-tailed ptarmigan female hobbling away from me. I followed her because I assumed that she was feigning injury. And she was. Suddenly, when I was about fifty meters from her young, she flew back to her chicks. I could never have followed her, and neither could any ground-dwelling predators such as red foxes or coyotes.

Dog-smarts and monkey-smarts

"Smart" and "intelligent" are loaded words and often are misused. To claim that variations in the behavior of different species are due to members of one species being less intelligent than members of another species shifts attention away from the various needs of the organisms that may explain the behavioral differences. Dogs are dog-smart and monkeys monkey-smart. Each does what is required to survive in its own world. Within species, it is likely that some animals are smarter or more adaptable than others, but variation in intelligence is to be expected as is variation in other traits. Individual variation provides the fuel for evolution. The misunderstanding and misapplication of the notions of smartness and intelligence can have significant and serious consequences for animals, especially if they are thought to be dumb and insensitive to pain and suffering.

There is no doubt that studying animal intelligence is challenging and exciting. There is still much to learn, and surprises are always forthcoming. Many animals are extremely clever, creative, and smart, perhaps even smarter than we will ever know! Our own intelligence will be challenged as we forge ahead to learn more about what other animals can do and what they know.

Some skeptics say that we can never know what another human, much less a member of another species, really is thinking or feeling, because she or he either could be lying to us or cannot tell us because of not having a "language." However, by using careful observations and some reasonable intuitions we can learn a lot about our kin. Not only can we learn about their behavior and what they think about, but we can also gather practical information concerning how humans influence other animals' lives when we intrude on them. We can also learn how we can make the lives of animals the best possible when they are held in captivity.

Animal consciousness and awareness

> To affirm, for example, that scallops 'are conscious of nothing,' that they get out of the way of potential predators without experiencing them as such and when they fail to do so, get eaten alive without (quite possibly) experiencing pain . . . is to leap the bounds of rigorous scholarship into a maze of unwarranted assumptions, mistaking human ignorance for human knowledge.
>
> —Maxine Sheets-Johnstone, "Consciousness"

I end this foray into animal minds by discussing the notion of animal consciousness. The brevity of this section is not because the topic is not of great interest and importance to researchers studying animal cognition and emotions but rather because there is a lot of controversy surrounding many of the difficult issues that need to be considered in discussions of animal consciousness. Some of the controversy turns on how "consciousness" is defined. We still do not have a good sense of what the words "consciousness," "self," and "self-identity" mean even for humans. Some disputes center on the fact that because we never can know with certainty what animals are thinking or if they are conscious, it is not worth studying this aspect of their behavior, for we will only be able to make better or worse guesses about their thoughts and conscious states (if any).

If "being conscious" means only that one is aware of one's surroundings, then many animals are obviously conscious. Simple awareness of this sort is called "perceptual consciousness."

It is when people make the jump from simple awareness to questions of self-consciousness or self-awareness (I will use these terms interchangeably for the sake of this discussion) in animals that the fur and feathers begin to fly. Many researchers argue that there are different degrees of consciousness. In addition to perceptual consciousness, there is also what some argue is a "higher" degree or level of consciousness, namely, self-consciousness, an awareness of who one is in the world. For example, as long as my brain works normally (and some of my friends would claim that this happens very infrequently, given that I do things like move "yellow snow" from one place to another), I know that I am Marc Bekoff, and I can be fairly sure that there are no other Marc Bekoffs who are exactly like me, with the same set of past experiences and future expectations. If something happens to me that I like or dislike, I know it is happening to me.

It is also possible that I may not know who I am but may be fully aware of something happening to my body. If I receive a blow to my head, I may not know my name or who I am, but I am able to feel the pain that is caused by my injury, and it would be wrong to make me suffer just because I do not know my name. This happened to me after a serious bicycling accident. I did not know who I was or where I was, but I experienced great pain because I had knocked out three of my front teeth and most of the skin was peeled off of the left side of my face as I slid across asphalt. My mother, Beatrice Rose, seems to be unaware of who she is but often responds to pleasurable and painful stimuli. There also is little doubt that many other animals know when something painful happens to their body.

Robert Mitchell notes that following Descartes, self-awareness is commonly viewed as a human characteristic not present in animals because they do not have language. Language and self-awareness, thought to be the most human of abilities, were of course first sought in great apes because of their closeness to human beings. It has been discovered in some "ape language" studies that some apes use personal pronouns, such as "me" and "you," and use language to describe their current circumstances, to describe their wants, and to plan activities, but they do not appear to use language to reflect on their past or present circumstances or to ponder ethical dilemmas.

The spot on the forehead: What does it mean?

While it is not known if other animals know *who* they are, some animals, especially chimpanzees and various monkeys, have been shown to use their mirror image to groom parts of their bodies that they cannot see—their teeth and their backs—without the mirror. Some chimpanzees, but not all, look into a mirror and touch a spot that was placed on their foreheads when they were sedated and thus unaware that the spot had been placed there. Some dolphins also respond to a spot on their foreheads as if they know that the spot is on themselves.

The ingenious "red spot" technique was first used by the psychologist Gordon Gallup. Some researchers argue that this sort of self-directed be-havior suggests not only that chimpanzees might have have a sense of their own bodies but also that their response to the red dot means to them "this is me," that they are self-conscious, they know "who" they are. (People in my mother's condition fail this test.) Whether this rich explanation for self-directed behaviors is warranted remains to be determined. But caution about what self-directed responses to the spot indicate should not be taken to mean that some animals do not know who they are. We just do not know very much about animal self-awareness as of now.

Moving beyond primates

The fascination with self-consciousness among ethologists and compara-tive scientists has largely been engendered by work on mirror self-recogni-tion in primates, but we need more research on other animals for whom the red spot experiments are not appropriate, for species in which touching the forehead is not a natural act. We need species-fair tests that, while designed by humans, take into account the natural behavior of the animals who are being studied. Failure to pass the spot test does not mean that an individual has no concept of self, and passing this test does not necessarily mean that an individual knows who he or she is.

As a working hypothesis and to broaden the array of animals in which self-awareness is investigated, I would not be surprised to find some rudi-ments of self-awareness in such highly social animals as wolves, carnivores who live in packs in which coordination and efficiency in communicating among individuals are essential for activities such as hunting, rearing young, and defending territory and food. It would be highly inefficient for an individual to have to guess all of the time what others know or what they are thinking or feeling. However, wolves could likely accomplish this by knowing that they are not another individual, that their body is not that of another, in the absence of knowing *who* they are. I also have often won-dered what information birds or other animals might accumulate about themselves by seeing their reflections in water.

Currently we do not know which animals are aware of themselves and which animals are not. Differences among species may be due to the use of a single method that is not suited for all species. We also do not know much about the sense of self in wild animals, for research on self-awareness has been done on a very limited number of individuals who have lived in close contact with humans, some of whom have required extensive training. While it may seem "obvious" to some humans that animals (for example, gorillas or dogs) who are closely related to us or with whom we are familiar are self-conscious, we really do not know for sure that this is so. What is very exciting is that research in this area is extremely challenging and leads to phenomenally interesting questions that might tell us about other aspects of behavior. Research in self-awareness will also inform arguments about human uniqueness.

The evolution of consciousness

Because I take a comparative and evolutionary approach to the study of behavior, I am interested in knowing if consciousness and self-consciousness serve some function that can be used to explain why these capacities might have evolved. In our book *Species of Mind*, Colin Allen and I were interested in knowing how ethology might contribute to the study of consciousness by focusing on questions about why consciousness has evolved. Our approach includes asking what functions consciousness might serve as animals deal with the demands of their social and nonsocial environments and whether or not a conscious organism might be better equipped to deal with environmental instability and complexity than a nonconscious one. Marian Dawkins points out that while it is difficult to know what to look for in studies of consciousness, this does not mean that it is impossible to make serious attempts to learn what characteristics are distinctive among animals who can be said to have conscious experiences. While Colin and I do not rule out the possibility that humans are the only individuals who are conscious, we believe that it is too soon to make such a judgment. Patrick Bateson correctly notes, "It would be as irresponsible as it would be illogical to suggest that because continuities might not be found, they do not exist."

Donald Griffin placed consciousness high on the agenda of cognitive ethology and suggested that consciousness evolved to allow adaptively flexible behavior. According to this suggestion, adaptively flexible behavior provides evidence of consciousness. It has also been suggested that consciousness evolved in social situations where it is important to be able to anticipate the flexible and adaptive behavior of others. If this is true, then complex social skills might be taken as evidence of consciousness. J. Piggins and C.J.C. Phillips argued that because there are energetic costs to the evolution of different degrees of awareness due to the neural apparatus on which awareness depends, animals who live in variable environments will evolve increased awareness, whereas those who live in more stable environments will not. Their view is consistent with that of others who see conscious awareness as being involved in behavioral flexibility. Their ideas lend themselves to empirical study, and in that light they are important for future studies of conscious awareness in animals.

Along with behavioral flexibility, features commonly cited to support attributions of consciousness include the integration of information from multisensory—visual, auditory, and olfactory—sources and language skills. Some researchers have assumed that consciousness provides an organism with a means to gain knowledge or information about its environment. If this is so, then perceptual capacities provide evidence of consciousness.

Behavioral flexibility and misrepresentation

Behavioral flexibility is relevant to attributions of consciousness because it is connected to an organism's monitoring of its own performance. An organism that cannot detect when its states misrepresent its environment will be much more limited with respect to the adjustments it can make when those states are caused by abnormal or unpredictable stimuli.

The extent to which various organisms can detect and respond to their own errors is an empirical question, and there is much room for matters of degree, for different varieties of mind. There are a number of areas that appear to be promising places to begin empirical studies. One such area is social play behavior, an activity that involves responding to behaviors that would elicit different responses in nonplay situations. Another area is social communication, particularly when signals are used deceptively or not performed in situations where they would be appropriate. Reactions such as surprise, embarrassment, and rapid learning (often involving just one or a few experiences with the conditions that caused the error) are all characteristics that might be shown by individuals who have knowledge of their own errors. For example, some animals appear surprised when their play signals are responded to with aggression; they seem to have expectations that play will follow.

Are we alone? Is there a need to know who one is?

If we pay attention to some basic and well-accepted biological ideas, especially evolutionary continuity, it is difficult to justify the belief that we are the only species on this planet in which individuals are self-conscious. But we still do not know if this is the case, and indeed we may be unique in our evolved capacity to know who we are and to use this ability in our daily activities. We do not know if other animals have a need to know who they are in the sense that most humans know who they are. We share with many animals common structures in the brain and also common neurotransmitters. Nevertheless, the notion of selfhood experienced by other animals is likely different from our own (as are their emotional feelings), if it exists at all. Charles Darwin argued that there are many connections among different animals, that there is *continuity* in evolution. Even if we are very different from dogs or cats, there is no reason to think that dogs, cats, or other animals are not conscious in their own species-specific ways. We really do not know the rudiments of selfhood. Indeed, it is possible, as Marc Hauser notes, that humans are unique in having the sense of themselves that they do, "to understand what it's like to have a sense of self."

Perhaps some animals simply do not need to know who they are. Piggins and Phillips postulate that "humans possess a significantly increased level of awareness, facilitated in particular by the acquisition of language, but that generally animals possess a level of awareness that is appropriate to their need." Thus, while it may be useful for humans to know themselves by name (and I sure find it useful to know who I am as "Marc"), Jethro likely does not have a need to know who he is to live a dog's life, and it is likely that this is so for many individuals of many other species. Animals do not have to write autobiographies. While individuals surely need to know that they are not another individual, this does not mean they need to be self-aware. Rather, it is necessary and sufficient only that they have a sense of their own bodies and body-awareness. Obviously, many animals are able to distinguish their own bodies—themselves—from others and represent themselves to themselves and to the world in this manner. Jethro knows that he is not Zeke, his buddy who visits him every day, and this knowledge is necessary and sufficient for Jethro to get on in the world of dogs.

I favor a view that recognizes different levels of consciousness and self-consciousness. However, I resist placing a higher value on animals who *seem* to be more self-aware than others. Knowing *who* you are is not necessarily "better" than knowing you are not another individual.

Looking for self in all the wrong places

It is obvious that many animals are awake and aware of their surroundings and that they respond to changes in social and nonsocial stimuli with which they are confronted simultaneously and sequentially. Many animals also behave "as if" they have a sense of self, but we do not really know this. What we do know is that many animals clearly behave in a manner that shows that they have some sense of their own bodies and that they behave as if their bodies are not the bodies of others. Whether body-awareness also indicates self-awareness for some species—that individuals know who they are—remains a mystery. Perhaps we are looking for self in all the wrong places or faces.

Getting beyond ourselves

Cognitive ethologists are concerned with the diversity and flexibility of solutions that individuals, even members of the same species, use for common problems and for problems that are unique to the species or to animals of different ages, gender, or social status. They stress broad comparisons and do not focus on a few select representatives of limited animal groups, such as white laboratory rats or pigeons. Studies of domestic animals have provided much useful information. What can be more fun and illuminating than to study your companion dog, cat, or fish?

While there has been a lot of focus on the cognitive abilities of nonhuman primates, there is also a wealth of data on nonprimates that is essential for broad comparative studies of animal cognition. The well-known ethologist Peter Marler, at the University of California, Davis, recently summarized his review of social cognition in nonhuman primates and birds as follows: "I am driven to conclude, at least provisionally, that there are more similarities than differences between birds and primates. Each taxon has significant advantages that the other lacks." Even the situation for nonhuman primates is not very clear-cut. Michael Tomasello and Josep Call, in their excellent book *Primate Cognition*, concluded that "the experimental foundation for claims that apes are 'more intelligent' than monkeys is not a solid one, and there are few if any naturalistic observations that would substantiate such broad-based, species-general claims."

Individual and species variation

Sweeping summaries about the cognitive abilities (or lack thereof) of members of a given species, while they can be important for answering questions about the evolution and ecology of cognition, can be misleading and ultimately have significant effects on how individuals are treated. A concentration on *individual* variation and not on species-level analyses is an important part of the agenda for future research in cognitive ethology. Often, people fail to note that generalizations about the cognitive skills of species are based on small data sets from a small number of individuals who may have been exposed to a narrow array of behavioral challenges. We know so little about so many species that sweeping generalizations need to be offered carefully. What we do not know might ultimately compromise the lives of many animals.

My own research on play behavior in wolves, coyotes, and dogs and antipredator behavior in various birds has taught me much about the worlds of other animals—how they think, how they solve problems, how they flexibly adjust their behavior to current demands, and how they feel about certain situations. By learning about the behavior and minds of other animals, it will be easier for humans to appreciate them for who they are. Barbara Smuts, a seasoned primatologist and author of *Sex and Friendship in Baboons,* reflects: "My own life has convinced me that the limitations most of us encounter in our relations with other animals reflect not their shortcomings, as we so often assume, but our own narrow views about who they are and the kinds of relationships we can have with them."

Where to from here?

Cognitive ethology is a young science, and we need not be apologists for the enormous gaps in our knowledge about others' minds. We need to be patient and give ethologists time to do the extremely challenging work that needs to be done. It is time to go out into the field to see what animals really are doing in the course of their days and nights. Methods of study need to be tailored to the animals and questions under consideration, and all hypotheses and explanations need to be considered. We do not yet have enough information for making reliable claims about the distribution of cognitive skills or theories of mind, such as that chimpanzees have theories of mind but monkey, dogs, and fish do not. Much research is also needed to identify those behavior patterns that are instances of consciousness and self-consciousness. Whether language is a necessary prerequisite for self-consciousness still remains debatable. Some people see this requirement as reserving self-consciousness and self-reflection only for humans.

five

◘

ANIMAL EMOTIONS

Passionate Natures and Animal Feelings

Flint and Flo

There is ample evidence that for many animals, especially vertebrates, the real question of interest is not *whether* they have emotional lives but rather *why* different emotions have evolved, what they are good for. To deny animals' emotions is to deny a large part of who these beings are.

> Never shall I forget watching as, three days after Flo's death, Flint climbed slowly into a tall tree near the stream. He walked along one of the branches, then stopped and stood motionless, staring down at an empty nest. After about two minutes he turned away and, with the movements of an old man, climbed down, walked a few steps, then lay, wide eyes staring ahead. The nest was one which he and Flo had shared a short while before Flo died. . . . In the presence of his big brother [Figan], [Flint] had seemed to shake off a little of his depression. But then he suddenly left the group and raced back to the place where Flo had died and there sank into ever deeper depression. . . . Flint became increasingly lethargic, refused food and, with his immune system thus weakened, fell sick. The last time I saw him alive, he was hollow-eyed, gaunt and utterly depressed, huddled in the vegetation close to where Flo had died. . . . The last short journey he made, pausing to rest every few feet, was to the very place where Flo's body had lain. There he stayed for several hours, sometimes staring and staring into the water. He struggled on a little further, then curled up—and never moved again.
> —Jane Goodall, *Through a Window*

Echo, Enid, and Ely: A mother's devotion

Cynthia Moss, who has studied the behavior of wild African elephants for more than three decades and who recently won a prestigious MacArthur fellowship for her unrelenting work on conserving these marvelous animals as director of the Amboseli Elephant Research Project in Kenya, tells the following story of a mother's devotion. The gestation period for elephants is twenty-two months, and a female gives birth to a single calf every four to five years. Mothers also lactate to provide food for about four years. In 1990, Dr. Moss made a film about a family of elephants called the EBs, whose leader, Echo, was a "beautiful matriarch." Echo gave birth in late February to a male, Ely, who could not stand up because his front legs were bent. Ely's carpal joints were rigid. Echo continuously tried to lift Ely by reaching her trunk under and around him. Once Ely stood he shuffled around on his knees for a short while and then collapsed to the ground.

When other clan members left, Echo and her nine-year-old daughter, Enid, stayed with Ely. Echo would not let Enid try to lift Ely. Eventually the three elephants moved to a water hole, and Echo and Enid splashed themselves and Ely. Despite the fact that Echo and Enid were hungry and thirsty, they would not leave an exhausted Ely. Echo and Enid then made low rumbling calls to the rest of their family. After three days, Ely finally was able to stand.

Ely is now twelve years old. Echo's devotion paid off. But there is more to this story, details of which could only be gathered by conducting long-term research on known individuals. When Ely was seven years old, he suffered a serious wound from a spear that was embedded about one foot into his back. Although Echo now had another calf, she remained strongly bonded to Ely and would not allow a team of veterinarians to tend to him. When Ely fell down after being tranquilized, Echo and other clan members tried to lift him. Echo, Enid, and another of Echo's daughters, Eliot, remained near Ely despite attempts by the veterinarians to disperse the elephants so that they could help Ely. The elephants refused to leave despite gunshots being fired over their heads. Finally, Ely was treated and survived the injury. Echo was there to attend to Ely when he was a newborn and later when he was juvenile. Moving elephants around by breaking up family groups to accommodate zoos and circuses is clearly unnatural.

Dr. Moss stresses that Echo's courage and passion epitomize all that is extraordinary about elephants. Joyce Poole, author of *Coming of Age with Elephants,* similarly notes: "It is hard to watch elephants' remarkable behavior during a family or bond group greeting ceremony, the birth of a new family member, a playful interaction, the mating of a relative, the rescue of a family member, or the arrival of a musth male, and not imagine that they feel very strong emotions which could be best described by words such as joy, happiness, love, feelings of friendship, exuberance, amusement, pleasure, compassion, relief, and respect."

Primate anger and empathy

In Tezpur, India, a troop of about a hundred rhesus monkeys brought traffic to a halt after a baby monkey was hit by a car. The monkeys encircled the injured infant, whose hind legs were crushed and who lay in the road unable to move, and blocked all traffic. A government official reported that the monkeys were angry, and a local shopkeeper said: "It was very emotional . . . Some of them massaged its legs. Finally, they left the scene carrying the injured baby with them."

In another incident, baboons in Saudi Arabia waited for three days on the side of a road to take revenge on a driver who had killed a member of their troop. The baboons lay in waiting and ambushed the driver after one baboon screamed when the driver passed by them. The angry baboons threw stones at the car and broke its windshield.

Captive Diana monkeys have been observed engaging in behavior that strongly suggests empathy. Individuals were trained to insert a token into a slot to obtain food. The oldest female in the group failed to learn how to do this. Her mate watched her failed attempts, and on three occasions he approached her, picked up the tokens she had dropped, inserted them into the machine, and then allowed her to have the food. The male apparently evaluated the situation, helped his mate only after she failed, and seemed to understand that she wanted food, but could not get it on her own. He could have eaten the food, but he let his mate have it. There was no evidence that the male's behavior benefited him in any way other than to help his mate.

Emotions as "social glue"

> Sometimes I read about someone saying with great authority that animals have no intentions and no feelings, and I wonder, "Doesn't this guy have a dog?"
> —Frans de Waal, interview, *New York Times*, 26 June 2001

It is an understatement to claim that people, especially researchers, disagree about the nature of animal emotions, especially the question of whether any animals other than humans can feel emotions. The ancient Greeks believed that many animals experience the same range of emotions as humans. Current research in ethology, neurobiology, endocrinology, psychology, and philosophy is providing compelling evidence that at least some animals feel a wide range of emotions, including fear, joy, happiness, shame, embarrassment, resentment, jealousy, rage, anger, love, pleasure, compassion, respect, relief, disgust, sadness, despair, and grief. Many people, including researchers, who are uncertain about whether some animals experience emotions do not hesitate to attribute emotions to companion dogs and cats in the absence of "hard data."

Some of the issues that I discussed in the previous chapter concerning the privacy of other minds surface once again, for just as we can never know for sure what other animals are thinking, we also can never know with certainty what they are feeling, other humans included. Furthermore, the nasty *A*-words, anthropomorphism and anecdote, once again raise their "ugly heads" (at least to some they are ugly). The worry that practicing anthropomorphism is fatal to studies of animal minds—animal thinking, consciousness, and emotions—is not justified.

There is little doubt emotions have a long evolutionary history, as do other behavioral traits. This is not to say that animal emotions are identical to ours, for even among humans joy, for example, feels different to different people, and humans also display fear, bereavement and grief, and anger in a wide variety of ways.

In *The Smile of a Dolphin* many well-respected researchers who have spent their lives studying and living with a wide variety of animals share their stories about the emotional lives of the animals they know best. Their stories, supported by empirical data, show that many animals experience deep emotions ranging from joyful glee when playing to bereavement, grief, and depression over the loss of a mate, child, or other friend. Their stories also show that scientists themselves have strong emotional feelings about the animals they study, despite the warnings of some of their colleagues that they should not get attached to their "research subjects," for it will taint their science.

Few nonresearchers doubt that many animals experience emotions. I was interviewed for an article about animal emotions ("Was fühlen Tiere") in the German magazine *Der Spiegel* that was stimulated by the massive slaughter of animals because of the threat of foot-and-mouth disease across the United Kingdom and continental Europe in spring 2001. The reporters wanted to know if the animals who were being killed suffered, and I told them, "Yes, they suffered twice, first when seeing, hearing, and smelling the slaughter of their friends and then when they themselves were killed." I also did an interview with the TV network Univision, because they wanted to know if dogs suffered when they were dropped alive into boiling water as they were being prepared for food in Korea. The answer again is a no-brainer. Of course they do. There is solid neurobiological data that show that many animals, including fish, possess pain receptors, known as noci-ceptors, and that individuals feel pain.

That the popular press is interested in animal emotions only reflects the deep concerns that many people have for animal feelings. During the foot-and-mouth crisis, in many of the interviews that I saw with farmers who had lived their entire lives with animals (as had their ancestors, often on the same land), families were crying, not merely because of economic losses but because they were asked to kill their friends, animals with whom they were closely bonded, beings they named and loved. Some people were crying be-cause they had to do the killing themselves, rather than having someone in a distant slaughterhouse do what spies call the "wet work." Seeing the ani-mals suffer, hearing them suffer, and smelling their pain reduced the dis-tance to a first-person experience that was intolerable. I will return to this theme when I write about animals as food, noting that many children do not know that a hamburger was a cow or the bacon on their bacon, lettuce, and tomato sandwich was a pig.

Shared emotions and empathy are the social glue for the development and maintenance of bonds with other animals. We can only tolerate so much alienation before the cycle reverses itself and some sort of reconnec-tion is sought. I often wonder if humans view animals as having some qual-ities that we have lost—their mindful presence, their unfiltered emotions, their zest for life. If animals, including those who are routinely used for re-search, education, amusement, food, and clothing, are aware of the emo-tional states of others, there are serious implications for considerations of their well-being. An additional dimension of awareness must be taken into account, because individuals enjoy and suffer not only their own but also others' feelings and fates. Added to concerns about how animals are treated by humans in captivity and in nature, considerations of animal emotions and empathy compound an already challenging and contentious debate about the humane treatment of animals, a topic to which I return in chap-ters 7 and 8.

Michael Tobias, a filmmaker, author, and ecologist, once found himself swimming with gigantic whale sharks and was taken by their gigantic heart, gentleness, and nonviolent nature. Their gentleness was so conta-gious that Tobias was "completely severed from time" and "unsnarled from all connections." He had become one with Nacho, the name given to one whale shark by a local doctor. The unfiltered emotions that exuded from Nacho were a strong glue for connecting Tobias with his newfound friend, for developing a trust and fellowship of mutual admiration. One reason that many animals are able to form close and reciprocal social bonds with one another (and with humans) is shared emotions.

What are emotions?

Emotions can be broadly defined as psychological phenomena that help in behavioral management and control. Yet no matter how hard people have tried to wrap up the concept of emotions into a tidy bundle, this has not been possible. Some people feel that the word "emotion" is so general that it escapes any single definition. Likewise, no single theory of emotions captures the complexity of the phenomena. Jaak Panksepp has suggested that emotions need to be defined in terms of their adaptive and integrative functions rather than their general input and output characteristics. I agree. It is important to extend our research beyond the underlying physiological mechanisms that mask the richness of the emotional lives of many animals and learn more about how emotions serve them as they go about their daily activities.

Generally, scientists and nonscientists alike seem to agree that emotions are real and that they are extremely important, at least to ourselves. While there is not much consensus on the nature of animal emotions, there is no shortage of views on the subject. Some people, following René Descartes and B. F. Skinner (inventor of the Skinner box), believe that animals are merely robots who become conditioned to respond automatically to stimuli to which they are exposed. The view of animals as machines explains so much about what they do that it is easy to understand why many people have adopted it.

However, not everyone accepts that animals are merely automatons, unfeeling creatures of habit. Why then are there competing views on the nature of animal emotions? In part, this is because some people view humans as unique and special animals. Humans, they say, were created in the image of God and are the only rational beings who are able to engage in self-reflection. The philosopher Bernard Rollin notes that at the end of the 1800s animals "lost their minds." In attempts to emulate the up-and-coming "hard sciences" such as physics and chemistry, researchers studying animal behavior came to realize that there was too little in studies of animal emotions and minds that was directly observable, measurable, and verifiable, so it was best to forget about them and to concentrate on behavior because overt actions could be seen, measured objectively, and verified.

Primary and secondary emotions

The emotional states of many animals, especially mammals, are easily recognizable. Their faces, their eyes, and the ways in which they carry themselves can be used to make strong inferences about what they are feeling. Changes in muscle tone, posture, gait, facial expression, eye size and gaze, vocalizations, and odors (pheromones), singly and together, indicate emotional responses to certain situations. Even people with little experience observing animals usually agree with one another on what an animal is most likely feeling. Their intuitions are borne out, because their characterizations of animal emotional states predict future behavior quite accurately.

Researchers usually recognize two different types of emotions, primary and secondary emotions. Primary emotions, considered to be basic inborn emotions, include generalized rapid, reflex-like ("automatic" or hard-wired) fear and fight-or-flight responses to stimuli that represent danger. Animals can perform a primary fear response such as avoiding an object, but they do not have to recognize the object generating this reaction. Loud raucous sounds, certain odors, and objects flying overhead often lead to an inborn avoidance reaction to all such stimuli that indicate "danger." Natural selection has resulted in innate reactions that are crucial to individual survival. There is little or no room for error when confronted with a dangerous stimulus.

Primary emotions are wired into the evolutionarily old limbic system (especially the amygdala), the "emotional" part of the brain, so named by Paul MacLean in 1952. Structures in the limbic system and similar emotional circuits are shared among many different species and provide a neural substrate for primary emotions. In his three-brain-in-one (triune brain) theory, MacLean suggested that there was the reptilian or primitive brain (possessed by fish, amphibians, reptiles, birds, and mammals), the limbic or paleomammalian brain (possessed by mammals), and the neocortical or "rational" neomammalian brain (possessed by a few mammals, such as primates) all packaged into the cranium. Each is connected to the other two, but each also has its own capacities. While the limbic system seems to be the main area of the brain in which many emotions reside, current research indicates that all emotions are not necessarily packaged into a single system, and there may be more than one emotional system in the brain.

Secondary emotions are those that are experienced or felt, those that are evaluated and reflected on. They involve higher brain centers in the cerebral cortex. Thought and action allow for flexibility of response in changing situations after evaluating which of a variety of actions would be the most appropriate to perform in the specific context. Although most emotional responses appear to be generated unconsciously, consciousness allows an individual to make connections between feelings and action and allows for variability and flexibility in behavior.

Emotion and cognition: How are emotions experienced by individuals?

Perhaps the most difficult of unanswered questions about animal emotions concerns how emotions and cognition are linked, how emotions are felt, or reflected on, by humans and other animals. In his book *The Feeling of What Happens: Body and Emotion in the Making of Consciousness*, Antonio Damasio provides a biological explanation for how emotions might be felt in humans. His explanation might also apply to some animals. Damasio suggests that various brain structures map both the organism and external objects to create what he calls a second-order representation. The mapping of the organism and the object most likely occurs in the thalamus and cingulate cortices. A sense of self in the act of knowing is created, and the individual knows "to whom this is happening." The "seer" and the "seen," the "thought" and the "thinker," are one in the same.

The evolution of emotions: Emotional fever in iguanas

Charles Darwin is usually credited with being the first scientist to give serious attention to the study of animal emotions. In his books *On the Origin of Species* (1859), *The Descent of Man and Selection in Relation to Sex* (1871), and *The Expression of the Emotions in Man and Animals* (1872), Darwin argued that there is continuity between humans and other animals in their emotional (and cognitive) lives, that there are transitional stages among species, not large gaps. The differences among many animals are differences in *degree* rather than differences in *kind*. In *The Descent of Man and Selection in Relation to Sex* Darwin claimed that "the lower animals, like man, manifestly feel pleasure and pain, happiness, and misery."

Concerning continuity, I quote the dean of ornithologists, Alexander Skutch, who at ninety-seven years of age was still conducting field research on birds in Costa Rica. In his delightful book *The Minds of Birds*, Skutch wrote:

> It is remarkable how often the sounds that birds make suggest the emotions that we might feel in similar circumstances: soft notes like lullabies while calmly warming their eggs or nestlings; mournful cries while helplessly watching an intruder at their nests; harsh or grating sounds while threatening or attacking an enemy. . . . Birds so frequently respond to events in tones such as we might use that we suspect their emotions are similar to our own.

An evolutionary and comparative approach to the study of emotions will help us learn more about the distribution of emotions among different species. Michel Cabanac has discovered that reptiles such as iguanas maximize sensory pleasure. He found that iguanas prefer to stay warm rather than venture out into the cold to get food, whereas amphibians such as frogs did not show such behavior. Neither did fish. Iguanas experience what is called "emotional fever" (a rise in body temperature) and tachycardia (increased heart rate), physiological responses that are associated with pleasure in other vertebrates, including humans. Cabanac postulated that the first mental event to emerge into consciousness was the ability of an individual to experience the sensations of pleasure or displeasure. Cabanac's research suggests that reptiles experience basic emotional states and that the ability to have an emotional life emerged between amphibians and early reptiles.

Naturalizing the study of animal emotions

Field research on animal emotions is of primary importance, for emotions have evolved in specific contexts. Naturalizing the study of animal emotions will provide for more reliable data than information collected in unnatural circumstances (although animals raised in extremely impoverished social environments display deep grief) because emotions have evolved just as have other behavioral patterns and organ systems, including stomachs, hearts, kidneys, and brains. Evolution is as private as the minds of other individuals. While we can make better or worse guesses about why a particular behavior pattern evolved, we cannot really know for sure that our reconstruction is *the* correct answer.

A lover's quarrel

The expression of emotions in animals raises many stimulating and challenging questions to which relatively little systematic empirical research has been devoted, especially in free-ranging individuals. Popular accounts (for example, Jeffrey Masson and Susan McCarthy's *When Elephants Weep*) have raised awareness of animal emotions, especially among nonscientists, and have provided scientists with much useful information for further systematic research. Such books also have raised hackles among many scientists for being "too soft," that is, too anecdotal, misleading, or sloppy. Yet *When Elephants Weep* was a landmark in bringing many issues to the table concerning the passionate lives of animals.

Stephen Jay Gould notes (in his foreword to *The Smile of a Dolphin*) that there is a lover's quarrel about the topic of animal emotions. Most people who know animals, including many researchers, believe (some would say they "know") that many animals have deep emotional lives, but it is very difficult to prove their existence with certainty. It is very difficult to state categorically that no other animals enjoy themselves when playing, are happy when reuniting, or become sad over the loss of a close friend. Consider wolves when they reunite, their tails wagging loosely to-and-fro and individuals whining and jumping about. Consider also elephants reuniting in a greeting celebration, flapping their ears and spinning about and emitting a vocalization known as a "greeting rumble." Likewise, think about what animals are feeling when they remove themselves from their social group, sulk at the death of a friend, stop eating, and die.

The skeptics speak: The "as if" disclaimer and privacy of other's minds

> Sure we can't predict or tell what anything [*sic*] is feeling, or what its response is anymore than we can another person, but most of their senses are far more developed than ours. They've been in the world a lot longer. They can tell what we're feeling, and I think we can tell how they're feeling. I mean I know. Anybody who's been around animals knows that you can and vice versa. It's madness to set up this artificial—for whatever perverted reason—to set up this artificial barrier between humans and animals, saying that one has no contact or ability to tell what the other one feels or what moods there are between species. It's the craziest thing I have ever heard of.
> —Rick Bass, "The Wild into the World"

Rick Bass is a very experienced and well-published naturalist who has lived closely with many animals for years, and he knows them well. His down-home common sense is appealing. In his book *The New Wolves*, Bass reports that a famous naturalist, Ernest Thompson Seton, once used the deep grief that a male wolf felt for the loss of his mate to trap and kill the bereaved male. Seton, who had been a wolf hunter before he became a writer, spread the scent of Blanca, the female, all over by dragging her body across trap lines and waited for her mate, Lobo, to return, only to be killed because he was looking for his missing and beloved mate. Surely we need more than just good tales, but these and other stories provide the raw material for more in-depth and careful studies of animal passions.

Nonetheless, there seem to be no avenues of inquiry or scientific data strong enough to convince some skeptics that other animals possess more than some basic primary emotions. One critic offered the following "argument," although it really is not an argument at all, in response to a reporter's question about whether he thought that animals experienced emotions (and I paraphrase): I believe that some animals have emotions, but it is very difficult to study emotions because we don't have the tools to do so, so therefore those who argue that animals feel their emotions are wrong.

Another skeptic, when asked if he believed animals experience emotions, did not answer the question but rather said that we can never know for sure if they do. While he is correct that we can never know for sure if animals *feel* their emotions, this does not tell us whether he believes that animals *experience* emotions.

I once listened to a colleague tell me how smart and emotional moose were. He knew, for he had hunted them in the past, and they were difficult to locate, track, and kill. He told me that they clearly were afraid and suffered from anxiety during the pursuit and experienced pain when they were shot if it was not a "clean shot." But in the very next sentence he claimed that camels were not sentient, they were dumb. I was incredulous and asked why camels are different from moose, and he said, "Oh, you know, they act stupid." I pointed out that being stupid does not mean that an individual cannot experience pain, and he shrugged his shoulders and walked off. Anne Dagg's stories of camels in *The Smile of a Dolphin* make it clear that they are both sentient and highly emotional.

Much progress is being made by many "hard" scientists including ethologists, psychologists, anthropologists, and neurobiologists. Shadow-boxing with close-minded skeptics is unlikely to result in making much headway, and we really do have the tools to study animal emotions. Many of the methods that we use are also used in other areas of behavioral research, and in all of these disciplines we are making much progress. Even if future research were to demonstrate that similar (or analogous) areas of a chimpanzee's or dog's brain showed the same activity as a human brain when a person reports being happy or sad, some skeptics would still hold tightly to the view that we simply cannot know what individuals are truly feeling, so these studies are fruitless. They claim that just because an animal acts "as if" she is happy or sad, we cannot say more than merely "as if," and such "as if" statements provide insufficient evidence.

Let me return to the concern about the privacy of other minds. The supposedly total privacy of other minds has stimulated the renowned evolutionary biologist George Williams to claim: I am inclined merely to delete it [the mental realm] from biological explanation, because it is an entirely private phenomenon, and biology must deal with the publicly demonstrable.

Using this quotation, Colin Allen and I laid out the following argument, which basically does away with cognitive ethology in the same manner that Cecelia Heyes dispensed with the field of research when she suggested that ethologists give up their field glasses, return to the laboratory, and have nothing more to say about animal consciousness. According to Williams:

(1) Mental events are private phenomena.
(2) Private phenomena cannot be studied biologically.

(3) Therefore, mental events cannot be studied biologically.

(4) Cognitive ethology is possible only if mental events can be studied biologically.

(5) Therefore, cognitive ethology is not possible.

If cognitive ethology is not possible, then the study of animal cognition, consciousness, and emotions is not possible, and we might as well close up shop and go home. Indeed, this is not what the future holds in store, but some people would be happy if this were the case. Williams's argument depends on a debatable philosophical premise (#1). It is ironic that these premises, which can only be defended in nonempirical, philosophical fashion, are produced by critics who would typically regard themselves as hard-nosed empiricists. We really do not know if other minds are *totally* private, and much research indicates that we can access much of what is in other minds and make very accurate predictions about future behavior using these inferences. Cognitive ethologists do empirical work, yet many of its critics rarely analyze the empirical work to determine whether it demonstrates what it is designed to show. Instead, they base their arguments on claims that are as fraught with interpretive difficulty as the cognitive conclusions they wish to deny.

Regardless of the privacy of other minds, we can still ask meaningful questions about the evolution of emotions. For example, if emotions clearly drive humans to action, why can it not be so that at least some other animals are also driven by their emotions? It is unlikely that secondary emotions evolved only in humans with no precursors in other animals. Joyce Poole notes, "While I feel confident that elephants feel some emotions that we do not, and vice versa, I also believe that we experience many emotions in common."

A potpourri of animal emotions

FEAR

Fear is one of the easiest emotions with which to deal. It is a primary emotion that is shared by many different animals, because "being afraid" is related to an individual's very survival. Often there are no second chances in nature; an animal has to perform the correct action the very first time he or she faces a dangerous situation, whether it be a predator or a stranger who is up to causing harm.

Animals faced with fear will usually cringe, run away, or face the situation squarely hoping to survive the encounter. Some animals, such as opossums, just freeze in place, to make themselves less conspicuous and perhaps hoping for the best.

It often is said that animals smell fear. Jethro enjoys going to the veterinarian, especially for acupuncture treatments for his sore elbow. But if the dog who had just been in the examination room was afraid, Jethro also shows hesitation and fear: he tucks his tail and pulls his ears back, rather than swinging his tail about (knocking things off the examination table) and holding his ears forward and alert. The smell of fear is conveyed by a glandular secretion from the anal gland of the previous and absent canine client. Rats who have been exposed to cats show fear responses when they are exposed to the odor of a cat. Evolution—natural selection—has resulted in inborn reactions that are crucial to individual survival. There is little or no room for error when confronted with a dangerous stimulus.

JOY, HAPPINESS, AND PLAY

Social play is an excellent example of a behavior in which many animals partake, and one that they enjoy immensely. There is an feeling of incredible freedom in the flow of play. Individuals become immersed in the activity, and there seems to be no goal other than to play; play is what is called an "autotelic" behavior.

Animals seek out play relentlessly, and when a potential partner does not respond to a play invitation, they often turn to another individual. If all potential partners refuse their invitation, individuals will play with objects or chase their own tails. Specific play signals are used to initiate and to maintain play. The play mood is contagious; just seeing animals playing can stimulate play in others. Animals seek out play because it is fun. Consider some of the notes I took while watching Jethro play with his buddy Zeke.

> Jethro runs towards Zeke, stops immediately in front of him, crouches or bows on his forelimbs, wags his tail, barks, and immediately lunges at him, bites his scruff and shakes his head rapidly from side-to-side, works his way around to his backside and mounts him, jumps off, does a rapid bow, lunges at his side and slams him with his hips, leaps up and bites him neck, and runs away. Zeke takes wild pursuit of Jethro and leaps on his back and bites his muzzle and then his scruff, and shakes his head rapidly from side-to-side. They then wrestle with one another and part, only for a few minutes. Jethro walks slowly over to Zeke, extends his paw toward Zeke's head, and nips at his ears. Zeke gets up and jumps on Jethro's back, bites him, and grasps him around his waist. They then fall to the ground and wrestle with their mouths. Then they chase one another and roll over and play.

It is more difficult to deny that Jethro and Zeke were having fun and enjoying themselves than to accept that they enjoyed what they were doing.

Studies of the chemistry of play support the idea that play is fun. Steve Siviy, working at Gettysburg College, has shown that dopamine (and perhaps serotonin and norepinephrine) is important in the regulation of play and that large regions of the brain are active during play. Rats show an increase in dopamine activity when anticipating the opportunity to play. Jaak Panksepp has found a close association between opiates and play and claims that rats enjoy being playfully tickled.

Siviy's and Panksepp's findings suggest that play is enjoyable. In light of these neurobiological ("hard") data concerning possible neurochemical bases for various moods, in this case joy and pleasure, skeptics who claim that animals do not feel emotions might be more likely to accept the idea that enjoyment could well be a motivator for play behavior.

GRIEF AND SORROW: THE EYES SAY IT ALL

"The light in their eyes simply goes out, and they die." So wrote Judy McConnery of traumatized orphaned gorillas. Many animals display grief at the loss or absence of a close friend or loved one. One vivid description of the expression of grief is offered at the beginning of this chapter, in Jane Goodall's description of observing Flint, an eight-and-a-half-year-old chimpanzee, withdraw from his group, stop feeding, and finally die after his mother, Flo, died.

The Nobel laureate Konrad Lorenz observed grief in geese that was similar to grief in young children. He provided the following account: "A greylag goose that has lost its partner shows all the symptoms that John Bowlby has described in young human children in his famous book *Infant Grief*. . . . The eyes sink deep into their sockets, and the individual has an overall drooping experience, literally letting the head hang."

Sea lion mothers watching their babies being eaten by killer whales squeal eerily and wail pitifully, in anguish of their loss. Dolphins have been observed struggling to save a dead infant. Barbara Smuts observed an African antelope, Pala, grieve the death of her infant who had been killed by olive baboons. Pala watched a baboon eat her infant, and then she chased the baboon away and gazed at the remaining skin and bones. Pala continued to stand motionless over her infant's body through the night. Perhaps Pala's silent vigil was her way of mourning her loss. Elephants have been observed to stand guard over a stillborn baby for days with their head and ears hung down, quiet and moving slowly as if they are depressed. Orphan elephants who saw their mothers being killed often wake up screaming. Iain Douglas-Hamilton, author of *Among the Elephants and Battle for the Elephants*, who has studied elephants for more than four decades, told me that he has no doubt that elephants behave differently to the bones from other elephants than they do to bones from individuals of other species.

LOVE IS A MANY SPLENDORED AND VARIED FEELING

Love appears in many forms among animals. Courtship and mating are two activities in which numerous animals regularly engage. Many animals seem to fall in love with one another just as humans do. In many species, romantic love slowly develops between potential mates. It is as if one or both need to prove their worth to the other before they consummate their relationship. Even fish act as if they love one another. Lee Dugatkin observed "guppy love." Males change their behavior and become bolder in response to a predator when there is a female around, because females find bold males more attractive. It seems like males will risk it all for love.

Bernd Würsig described courtship in southern right whales off Peninsula Valdis, Argentina. While courting, Aphro (female) and Butch (male) continuously touched flippers, began a slow caressing motion with them, rolled toward each other, briefly locked both sets of flippers as in a hug, and then rolled back up, lying side by side. They then swam off, side by side, touching, surfacing and diving in unison. Würsig followed Butch and Aphro for about an hour, during which they continued their tight travel. Würsig believes that Aphro and Butch became powerfully attracted to each other and had at least a feeling of "afterglow" as they swam off. He asks, could this not be leviathan love? Aphro and Butch's behavior resembles activities that we call "love" in other animals with whom we are more familiar.

Many things have passed for love in humans, yet we do not deny its existence, nor are we hesitant to say that humans are capable of falling in love. It is unlikely that romantic love (or any emotion) first appeared in humans with no evolutionary precursors in animals. Indeed, there are common brain systems and homologous chemicals underlying love (and other emotions) that are shared among humans and animals. The presence of these neural pathways suggests that if humans can feel romantic love, then at least some other animals also experience this emotion.

LOVE IN DOGS

We all know that dogs can be our very best friends. They can also be one another's best friend. Two malamutes, Tika and her mate, Kobuk, had raised eight litters of puppies together, and now they were enjoying their retirement years living with Anne Bekoff. Kobuk was charming and energetic, always demanding attention. He would always let you know when he wanted his belly rubbed or his ears scratched. He also was quite vocal and howled his way into everyone's heart. Tika was quieter and pretty low-key. If anyone tried to rub Tika's ears or belly, Kobuk shoved his way in. Tika knew not to eat her food unless it was far away from Kobuk. If Tika happened to get in Kobuk's way when he headed to the door, she usually got knocked over as he charged past her. Tika and Kobuk learned to work out their differences over the years that they were together.

But things between Kobuk and Tika were about to change. One day a small lump appeared on Tika's leg. It was diagnosed as a malignant tumor. Overnight Kobuk's behavior changed. He became subdued and would not leave Tika's side. After she had her leg amputated and was having trouble getting around, Kobuk got worried about her. He stopped shoving her aside, and he did not even mind if she was allowed to get on the bed without him.

About two weeks after Tika's surgery, Kobuk woke Anne up in the middle of the night the way he does when he really needs to go outside. Tika was in another room, and Kobuk ran over to her. Anne got her up too and took them outside, but both of them just lay down on the grass. Anne heard Tika whining softly. She saw that Tika's belly was huge and swollen. Tika was going into shock, but the veterinarian operated on her and was able to save her life.

If Kobuk had not fetched Anne, Tika almost certainly would have died. Tika recovered, and Kobuk became the bossy dog he had been, even as Tika walked around on three legs. But now Anne knew that Kobuk would be there for Tika if she needed him. Tika knew this all the time.

ANGER AND AGGRESSION: BEWARE THE DOLPHIN'S SMILE

Animals can get angry just as humans get angry. And their anger is raw and unfiltered. Sometimes we do not really know that the animals are angry with one another, but it certainly looks as if they are; they threaten one another and often fight until one individual submits, leaves, or is harmed or killed. Sue Savage-Rumbaugh observed sibling rivalry between star-struck Kanzi and his younger sister, Panbanisha, when the *London Times* opted to take a picture of her rather than of him.

It is important to study animals carefully. Dolphins almost always look as if they are smiling, and people are likely to think, "Oh, they're so cute—they're so happy and friendly." But as Toni Frohoff, who has studied dolphins for many years, stresses that this can be a dangerous assumption because the dolphin's smile is not always what it appears to be. The smile usually means that the animal is upset or angry and should be avoided. Obviously, misreading the dolphin's smile can be risky. This is not to say dolphins cannot be friendly and loving, but the truth behind the dolphin's smile does caution us that the emotional lives of dolphins and many other animals can be complex. We should not make the animals who we want them to be, but rather we should appreciate them for who they are and for their own ways of expressing themselves and living.

Dolphin anger has been observed by Denise Herzing. When spotted dolphins get angry, the whites of their eyes become more visible, they arch their bodies, they swim more erratically, they open their mouths, and they emit barks and squawks. Spotted dolphins often encounter bottlenose dolphins, who are hundreds of pounds heavier and about three feet longer. It takes a group of spotted dolphins to fend off a bottlenose dolphin. Stubby, a spotted dolphin, once found himself being mounted by two male bottlenose dolphins, Sly and Max. In the absence of his group mates, Punchy, Big Wave, and Knuckles, Stubby was passive and allowed Sly and Max to harass him. But at a later time, when Punchy, Big Wave, and Knuckles were with him, Stubby and the other spotted dolphins pursued and fought with Sly. Stubby's eyes were white, he vocalized loudly, and his anger toward Sly was obvious and persistent.

Anger can also be displayed gently. Giraffes will often "neck" one another when they are upset. Necking involves the giraffes walking toward one another and standing side by side. Sometimes one male will gently graze the other male's neck with his horns. Anne Dagg, who studied giraffes for over twenty years, never saw necking lead to a real fight in male giraffes.

Even octopuses seem to get angry. Their pearly white skin turns red when they are agitated. Roland Anderson, who studies octopuses at the Seattle Aquarium, claims that octopuses wear their hearts on their skin for all to see. A red octopus is likely an angry octopus who should be avoided.

Birds also can get angry. Bernd Heinrich once experienced the anger of a raven when he blocked his way to food. Ravens also get angry with one another when one raids a food source. Irene Pepperberg, who has studied Alex, a gray parrot, noted that when Alex got frustrated because something that he expected did not happen, he seemed to get very angry. Alex would narrow his eyes, puff up his feathers, and lower his head when he was fed a pellet of bird food rather than a much preferred cashew.

A FROSTY HOMECOMING IN HYENAS

Spotted hyenas are African carnivores who live in clans. These efficient hunters feed mainly on large antelope they kill themselves. While they can get angry and fight with one another, usually they work out their differences without fighting. If one hyena is bothered by one of its clan mates, all it has to do is approach the other hyena or wave its head toward the other hyena. Usually the other hyena walks away, but if this doesn't happen there may be a chase. Biting seldom occurs, and only when the hyenas are really mad at one another.

Kay Holekamp and Laura Smale, who work out of Michigan State University, studied spotted hyenas in Kenya. They saw intense aggression only once, when an adult female hyena named Little Gullwing attempted to re-join her clan after being away for a year. Little Gullwing's former friends treated her as a trespasser, and they were very nasty to her. They were so aggressive because they thought that she would try to get food, and there was not enough to feed another mouth. Little Gullwing was being treated by her old friends as if she were a member of another clan.

Little Gullwing was nervous about rejoining her old friends. She approached them with her ears flattened against her head, her tail between her legs, and wearing a grin to calm her friends. As she approached she bobbed her head up and down and side to side to tell them that she wanted peace and only wanted to be their friend.

But her old friends did not accept her, and they raised their tails and put their ears forward while the hair on their backs stood up. They were excited and not happy about Little Gullwing's return. In response Little Gullwing dropped to the ground and crawled toward the other hyenas. Crawling usually stops other hyenas from being aggressive, but this time it did not work.

Little Gullwing was not welcomed. She was bitten and screamed in pain. After a number of hyenas bit her and drew blood, Little Gullwing got up and ran away from her former friends, who were now very angry. They chased her until she went into the territory of another clan of hyenas, and then they relaxed. Finally, Little Gullwing was accepted back. But it had not been an easy homecoming.

Embarrassment: I didn't do that

Jane Goodall observed what could be called embarrassment in chimpanzees. When Fifi's oldest child, Freud, was five and a half years old, his uncle, Fifi's brother Figan, was the alpha male of the chimpanzee community. Freud always followed Figan; he hero-worshipped the big male. Once, as Fifi groomed Figan, Freud climbed up the thin stem of a wild plantain. When he reached the leafy crown, he began swaying wildly back and forth. Had he been a human child, we would have said he was showing off. Suddenly the stem broke and Freud tumbled into the long grass. He was not hurt. He landed close to Jane, and as his head emerged from the grass, she saw him look over at Figan. Had he noticed? If he had, he paid no attention but went on grooming. Freud very quietly climbed another tree and began to feed.

Marc Hauser, of Harvard University, observed what could be called embarrassment in a male rhesus monkey. After mating with a female, the male strutted away and accidentally fell into a ditch. He stood up and quickly looked around. After sensing that no other monkeys saw him tumble, he marched off, back high, head and tail up, as if nothing had happened.

Where to from here?

The best way to learn about the emotional lives of animals is to conduct comparative and evolutionary ethological, neurobiological, and endocrinological research and to resist claims that anthropomorphism has no place in these efforts. To claim that one cannot understand elephants, dolphins, or other animals unless "we are one of them" leaves us nowhere. It is important to try to learn how animals live in their own worlds, to understand their own perspectives. Animals evolved in specific and unique situations, and it discounts their lives if we only try to understand them from our own perspective. Certainly, gathering information on animal emotions is difficult, but it is not impossible.

Perhaps so little headway has been made in the study of animal emotions because of a fear of being "nonscientific." In response to my invitation to contribute an essay to my book on animal emotions, one colleague wrote: "I'm not sure what I can produce, but it certainly won't be scientific. And I'm just not sure what I can say. I've not studied animals in natural circumstances and, though interested in emotions, I've 'noticed' few. Let me think about this." On the other hand, many other scientists were very eager to contribute. They believed we can be scientific and at the same time use other types of data to learn about animal emotions. It is permissible for scientists to write about matters of the heart.

Meeting the devil

Jaak Panksepp, in his book *Affective Neuroscience*, provides a useful thought experiment. Imagine that you are faced with making a devil's choice concerning the existence of animal emotions. You must answer correctly the question of whether or not other mammals have internally experienced emotional feelings. If you give the wrong answer you will follow the devil home. In other words, the stakes are high. Panksepp asks how many scientists would deny under these circumstances that at least some animals have feelings. Likely, few.

Leaving the tower for the field

It is important that researchers in various fields—ethology, neurobiology, endocrinology, psychology, and philosophy—coordinate their efforts to learn about animal emotions. No single discipline will be able to answer all of the important questions that still need to be dealt with. Laboratorybound scientists, field researchers, and philosophers must share data and ideas. A few biologists have entered into serious dialogue with philosophers, and some philosophers have engaged in field work. Colin Allen, a philosopher, spent a year with me learning how to conduct field work on birds and saw firsthand what a bird's-eye view of the world might look like. The philosopher Daniel Dennett, who has written much about cognitive ethology and animal minds, spent time watching vervet monkeys with Dorothy Cheney and Robert Seyfarth, authors of *How Monkeys See the World*. As a result of these collaborations, ethologists and philosophers can experience one another's views and come to an understanding of how the science and theorizing are done.

Future research must focus on a broad array of species and not on those animals with whom we are familiar (for example, companion animals) or those with whom we are closely related (nonhuman primates), animals to whom many of us freely attribute secondary emotions and a wide variety of moods. Species differences in the expression of emotions and what they feel like also need to be taken into account. Even if joy and grief in dogs are not the same as joy and grief in chimpanzees, elephants, or humans, this does not mean that there is no such thing as dog-joy, dog-grief, chimpanzee-joy, or elephant-grief. Even wild animals (for example, wolves), and their domesticated relatives (dogs), may differ in the nature of their emotional lives.

By remaining open to the idea that many animals have rich emotional lives, even if we are wrong in some cases, little truly is lost. By closing the door on the possibility that many animals have rich emotional lives, even if they are very different from our own or from those of animals with whom we are most familiar, we will lose great opportunities to learn about their lives.

There are many worlds beyond human experience. There are no substitutes for listening to, and having direct experiences with, other animals. Do animals love one another? Do they mourn the loss of friends and loved ones? Do they resent others? Can they be embarrassed? If we ask such questions, certainly our own lives will be richer for the effort, and the lives of other animals more understood, appreciated, and respected.

six

PLAY, COOPERATION, AND THE EVOLUTION OF SOCIAL MORALITY

Foundations of Fairness

I now want to discuss social play behavior in more depth and argue not only that animals enjoy playing but that during play they learn codes of social conduct that influence how they interact with other animals. Thus play may be a "foundation of fairness." While I will not discuss human play, I am reminded of Stuart Brown's observation that one of the best predictors of extreme human sociopathological behavior is a lack of social play when an individual is a youngster. Whether or not sociopaths did not learn how to behave fairly remains an open question, but surely their moral development suffered from their not engaging in interactive play during childhood.

Playing is fun, and so is learning about the nitty-gritty details and rhythms of play, an appreciation and understanding of which may help us learn about the evolution of social morality. I will discuss some of my research on members of the dog family to make the case for the possible existence of "moral mutts" and show that we may not have to turn to our next of kin, nonhuman primates, to seek out the roots of human egalitarianism and morality. Some people argue that humans are alone in this arena, but I am not so sure. Much more comparative and evolutionary research is needed before such speciesistic claims can be accepted carte blanche.

Science, theology, cooperation, and social morality

There are many areas in which scientists can pursue interesting and important questions that center on human spirituality and the place of humans in the world. One such area concerns the evolution of social morality. Many people often wonder if some animals have codes of social conduct that regulate their behavior in terms of what is permissible and what is not permissible during social encounters. They want to know just what the moral capacities of animals are. Are they moral agents with a moral sense who are able to live in moral communities? People also are interested in discussing the animal foundations on which human morality might be built, even if it is not identical to animal morality. Darwin's ideas about evolutionary continuity, that behavioral, cognitive, emotional, and moral variations among different species are differences in *degree* rather than difference in *kind*, are often invoked in such exercises. This view argues that there are shades of gray among different animals and between nonhumans and humans, that there is not a void in the evolution of moral capacity or agency.

The study of the evolution of morality, specifically cooperation and fairness, is closely linked to science, religion, theology, spirituality, and perhaps even different notions of God, in that ideas about continuity and discontinuity (the possible uniqueness of humans and other species), individuality, and freedom need to be considered in detail.

Recently, Gregory Peterson speculated on the evolutionary roots of morality (stages that he refers to as "quasi-morality" and "proto-morality" in animals) and religion in relation to the roles played by cognition and culture. Peterson stressed the importance of recognizing continuities and discontinuities with other animals, arguing that while some animals might possess proto-morality because they are able "to rationally deliberate actions and their consequences," none other than humans is "genuinely moral," because to be able to be genuinely moral requires higher emergent levels of cognition as well as culture and the worldview that culture provides, namely, religion. Peterson asserts that "quasi-moral and proto-moral systems do not require a global framework that guides decision making. They are always proximate and pragmatic. In these systems, there is no long-term goal or ideal state to be achieved. Yet, genuine morality is virtually inconceivable without such conceptions." Peterson also claims that any sociobiological account (based on selfishness or combativeness) of human morality is incomplete. This is so for some nonhuman animals as well.

My perspective—shared with others—for coming to an understanding and appreciation of animal behavior centers on the fact that in their own worlds animals may indeed have their own form of genuine morality, and there might indeed be long-term goals and ideal states to be achieved. Our anthropocentric view of other animals, in which humans are so taken with themselves to the exclusion of other animals, is far too narrow. Animals must be studied in their own worlds. The worlds and lives of other animals are not identical to those of humans and may vary from species to species and even within species. There is also variability among humans in what some might view as long-term goals and ideal states, and it would of course be premature to conclude that there is one set of long-term goals and ideal states that characterize, or are essential to, the capacity to be genuinely moral. We really are not experts about ourselves.

The evolution of social morality: Behaving fairly and cognitive empathy

Evolutionary reconstructions of social behavior often depend on educated guesses (some better than others) about the past social (and other) environments in which ancestral beings lived. Often it is difficult to know with a great deal of certainty very much about these variables and how they may have figured into evolutionary scenarios. It is extremely difficult to study the evolution of morality in any animal species, and the very notion of animal morality itself often makes for heated discussions. Irwin Bernstein's concern that "morality in animals might lie outside of the realm of measurement techniques available to science" needs to be taken seriously. Nonetheless, it seems clear that detailed comparative analyses of social behavior in animals can provide insights into the evolution of social morality. These sorts of studies are extremely challenging, but the knowledge that is gained is essential in our efforts to learn more about the evolution of sociality and social morality, as well as about human nature and perhaps human uniqueness.

Here I am specifically concerned with the notion of "behaving fairly," using as a working guide the notion that animals often have social expectations when they engage in various sorts of social encounters the violation of which constitutes being treated unfairly because of a lapse in social etiquette. It is through social cooperation that groups (communities) are built from individuals agreeing to work in harmony with other individuals. Whether or not individuals lose various "freedoms" when balanced against the benefits that accrue when they work for the "good of a group" is unknown and needs to be studied more carefully in various species. My view also emphasizes that cooperation is not always merely a by-product of tempering aggressive and selfish tendencies (combating selfish genes) and attempts at reconciliation. Rather, cooperation and fairness can evolve on their own because they are important in the formation and maintenance of social relationships. This view, in which nature is sanitized, contrasts with those who see aggression, cheating, selfishness, and perhaps amorality as driving the evolution of sociality. The combative Hobbesian world in which individuals are constantly at one another's throats is not the natural state of affairs, nature is not always red in tooth and claw, and altruism is not always simply selfishness disguised.

In his book *Good-Natured: The Origins of Right and Wrong in Humans and Other Animals*, Frans de Waal discusses the notion of *cognitive empathy*, "the ability to picture oneself in the position of another individual." De Waal claims that cognitive empathy may not be widespread among animals and "may be absent in other animals" except for people and perhaps the great apes. And it may not be absent. De Waal is correctly hesitant in drawing his conclusions. Only additional research will inform us about the distribution of cognitive empathy in different species.

Does it feel good to be fair?

Is it possible that it feels good to be nice to others, to cooperate with them and to treat them fairly, to forgive them for their mistakes and shortcomings? In the 1960s a study was done by Stanley Wechlin and his colleagues that showed that a hungry rhesus monkey would not take food if taking food subjected another monkey to electric shock. While no one knows what the hungry monkey was feeling about the mental state of the other monkey, it would be interesting to follow up on this study in more detail using more ethically acceptable methods including neuroimaging. There have also been studies that show that rats will sacrifice something when their actions to achieve a goal cause pain to another individual.

The importance of studying social carnivores

While many researchers focus on nonhuman primates as the most likely animals to show precursors to human morality, others have argued that we might learn as much or more about the evolution of human social behavior by studying social carnivores, species whose social behavior and organization resemble that of early hominids in a number of ways (divisions of labor, food sharing, care of young, and intersexual and intrasexual dominance hierarchies). We need long-term field studies of social animals for which it would be reasonable to hypothesize that emotions and morality have played a role in the evolution of sociality—that emotions and morality are important in the development and maintenance of social bonds that allow individuals to work together fairly and cooperatively for the benefit of all group members.

Animal play

"Happiness is never better exhibited than by young animals, such as puppies, kittens, lambs, &c., when playing together, like our own children." So

Three dogs at play. These dogs were studied on the campus of Washington University, St. Louis, Missouri, as part of my doctoral dissertation.

wrote Charles Darwin in his book *The Descent of Man and Selection in Relation to Sex.*

Animal play is obvious, but animal social morality is not. Social play in animals is an exhilarating activity to observe. The rhythm, dance, and spirit of animals at play is incredibly contagious. Not only do their animal friends want to join in or find others with whom to romp, but I also want to play when I see animals chasing one another, playing hide-and-seek, and wrestling with reckless abandon. Birds playfully soar across the sky chasing, diving here and there, and frolicking with one another. The unmistakable emotions associated with play—joy and happiness—drive animals into becoming at one with the activity. One way to get animals (including humans) to do something is to make it fun, and there is no doubt that animals enjoy playing.

How do animals tell others "I want to play with you"?

When individuals play, they typically use actions that are also used in other contexts, such as predatory behavior, antipredatory behavior, and mating. These actions may not vary much across different contexts, or they may be hard to discriminate even for the participants. I have long been interested in the following questions: How do animals know that they are playing?

How do they communicate their desires or intentions to play or to continue to play? How is the play mood maintained?

Because there is a chance that various behavior patterns that are performed during ongoing social play can be misinterpreted, individuals need to tell others "I want to play," "This is still play no matter what I am going to do to you," or "This is still play regardless of what I just did to you." An agreement to play, rather than to fight, mate, or engage in predatory activities, can be negotiated in various ways. Individuals may use various behavior patterns, called "play markers," to initiate play or to maintain a play mood by punctuating play sequences with these actions when it is likely that a particular behavior may have been, or will be, misinterpreted. (It is also possible that there are auditory, olfactory, and tactile play markers.)

After years of studying play in infant canids, I found that a play signal called the "bow" was used nonrandomly, especially when biting accompanied by rapid side-to-side shaking of the head was performed during play. The bow is easily recognizable and occurs when a dog crouches on her forelimbs while keeping her hind end upright. Barking and tail-wagging often accompany a bow. Bows are performed most frequently immediately before and immediately after biting accompanied by rapid side-to-side shaking of the head, an action that is performed during serious aggressive and predatory encounters and can easily be misinterpreted if its meaning is not modified by a play signal. Used in this way, a bow essentially serves as an exclamation point (!), telling the animal who is bitten that the bites are not meant to be taken seriously.

There is little evidence that play signals are used to deceive others in canids or other species. Cheaters are unlikely to be chosen as play partners, because others can simply refuse to play with them (I thank David Sloan Wilson for making this point) and choose others. Some of my own observations on infant coyotes show that cheaters have difficulty getting other young coyotes to play. It is not known if individuals select play partners based on what they have observed during play by others.

Dogs, wolves, and coyotes also engage in role-reversing and self-handicapping to maintain social play. Each can serve to reduce inequalities in size and dominance rank between the interacting animals and promote the reciprocity that is needed for play to occur. Self-handicapping (or "play inhibition") happens when an individual performs a behavior pattern that might compromise herself. For example, a coyote might not bite her play partner as hard as she can, or she might not play as vigorously as she can. Inhibiting the intensity of a bite during play likely helps to maintain the play mood. I once picked up a twenty-two-day-old coyote only to have him bite through my thumb with his needle-sharp teeth. His bite drew blood, and it really hurt. The fur of young coyotes is very thin, and an intense bite results in high-pitched squeals and much pain to the recipient. In adult wolves, a bite can generate as much as 1,500 pounds of pressure per square inch, a good reason to inhibit its force. Red-neck wallabies also engage in self-handicapping. They adjust their play to the age of their partner. When a partner is younger, the older animal adopts a defensive, flat-footed posture, and pawing rather than sparring occurs. In addition, the older player is more tolerant of its partner's tactics and takes the initiative in prolonging interactions.

Role-reversing occurs when a dominant animal performs an action during play that would not normally occur during real aggression. For example, a dominant animal might not voluntarily roll over on his back during fighting but would do so while playing. In some instances role-reversing and self-handicapping might occur together. For example, a dominant individual might roll over while playing with a subordinate animal and inhibit the intensity of a bite. From a functional perspective, self-handicapping and role-reversing, similar to using specific play invitation signals, might serve to signal an individual's intention to continue to play.

Why cooperate and play fairly?

For years I tried to figure out why play evolved as it did. Why do animals carefully use play signals to tell others that they really want to play and not try to dominate them? Why do they engage in self-handicapping and role-reversing? One morning, while hiking with Jethro, I had one of those infamous "aha" experiences and the puzzle was solved. It dawned on me that during social play, while individuals are having fun in a relatively safe environment, they learn ground rules about behavior patterns that are acceptable to others—how hard they can bite, how roughly they can interact—and how to resolve conflicts. There is a premium on playing fairly and trusting others to do so as well. There are codes of social conduct that regulate actions that are and are not permissible, and the existence of these codes likely speaks to the evolution of social morality. What could be a better atmosphere in which to learn social skills than during social play, where there are few penalties for transgressions? Individuals might also generalize codes of conduct learned in playing with specific individuals to other group members and to other situations such as sharing food, defending resources, grooming, and giving care. (Social morality does not mean other animals are behaving unfairly when they kill for food, for example, for they have evolved to do this.)

Playtime generally is safe time: transgressions and mistakes are forgiven and apologies are accepted, especially when one player is a youngster who is not yet a competitor for social status, food, or mates. There is a certain innocence or ingenuousness in play. Individuals must cooperate with one another when they play—they must negotiate agreements to play. Play is a voluntary activity. Robert Fagen, who has studied play for many years, noted that "levels of cooperation in play of juvenile primates may exceed those predicted by simple evolutionary arguments." The highly cooperative nature of play has evolved in many other species. Detailed studies of play in various species indicate that individuals trust others to maintain the rules of the game.

While there have been numerous discussions of cooperative behavior in animals, none has considered social play—the requirement for cooperation and reciprocity—and its possible role in the evolution of social morality, namely, behaving fairly. Individuals of different species seem to fine-tune ongoing play sequences to maintain a play mood and to prevent play from escalating into real aggression. Detailed analyses of film show that in canids there are subtle and fleeting movements and rapid exchanges of eye contact that suggest that players are exchanging information on the run, from moment to moment, to make certain everything is all right—that this is still play. Owen Aldis, in his book *Play Fighting*, suggested that in play there is a fifty-fifty rule so that each player "wins" about 50 percent of his or her play bouts, adjusting behavior to accomplish this. This seems to be the case for rats, as discovered by Sergio Pellis at the University of Lethbridge. Sequences consisting of assessing, monitoring, fine-tuning, and changing one's behavior to maintain play suggest that fairness and trust are important in ongoing play interactions.

Why might animals fine-tune play? Play in most species does not take up much time and energy, and in some species only minimal amounts of social play during short windows of time early in development are necessary to produce socialized individuals. For domestic dogs, two twenty-minute play sessions twice a week, between three and seven weeks of age, are sufficient for proper socialization. Researchers agree that play is very important in social, cognitive, and/or physical development. Marek Spinka, Ruth Newberry, and I have argued that play may also be important for training youngsters for unexpected circumstances. While there are few data concerning the actual benefits of social play in terms of survival and reproductive success, it generally is assumed that short-term and long-term functions (benefits) vary from species to species and among different age groups and between the sexes within a species.

No matter what the functions of play may be, there seems to be little doubt that the absence of play can have devastating effects on social development. In many species in which the young are dependent on older caregivers, there is a small time window during early development when individuals can play without being responsible for their own well-being.

This time period is generally referred to as the "socialization period," for this is when species-typical social skills are learned most rapidly. During the socialization period it is important for individuals to engage in at least *some* play. There is a premium for playing fairly if one is to be able to play at all. If individuals do not play fairly, they may not be able to find willing play partners. In coyotes, for example, youngsters are hesitant to play with an individual whom does not play fairly or with an individual who they fear. In many species individuals also show play-partner preferences, and it is possible that these preferences are based on the trust that individuals place in one another.

Neurobiological bases of sharing intentions and mind-reading

How might a play bow (or other action) serve to provide information to its recipient about the sender's intentions? Is there a relationship among acting, feeling, seeing, and feeling/knowing? Perhaps one's own experiences with play can promote learning about the intentions of others. Perhaps the recipient of a bow shares the intentions (beliefs, desires) of the sender based on the recipient's own prior experiences of situations in which she performed play bows. Recent research by Vittorio Gallese and his colleagues at the University of Parma in Italy suggests a neurobiological basis for sharing intentions. "Mirror neurons," found in the cerebral (premotor) cortex of macaques, fire when a monkey performs an action and also when the monkey observes the *same* action performed by another monkey. Gallese believes that mirror neurons might also be used in other modalities such as audition and olfaction.

Research on mirror neurons is truly exciting, and the results of these efforts will be very helpful for answering questions about which species of animals may have "theories of mind" or "cognitive empathy" about the mental and emotional states of others. Gallese and the philosopher Alvin Goldman suggest that mirror neurons might "enable an organism to detect certain mental states of observed conspecifics . . . as part of, or a precursor to, a more general mind-reading ability." Laurie Carr and her colleagues at the University of California at Los Angeles discovered, by using neuroimaging in humans, similar patterns of neural activation both when an individual observed a facial expression depicting an emotion and when they imitated the facial expression. This research suggests a neurobiological underpinning of empathy. Chris and Uta Frith have also reported results of neural imaging studies in humans that suggest a neural basis for one form of "social intelligence," understanding others' mental states (mental state attribution).

Finally, it has been reported by Josie Glausiusz that the neurobiologists Jean Decety and Perrine Ruby, working at the French Institute of Health and Medical Research in Lyons, have collected data that suggest that empathy is hard-wired in the human brain. Decety and Ruby asked ten young males to imagine such common actions as peeling a banana while undergoing a PET scan. The subjects were then asked to think about another person performing the same task. In both situations the PET scan revealed that the part of the brain that controls voluntary muscle control was active. But when the subjects were imagining someone else peeling a banana, the parietal cortex in the right hemisphere of the brain also lit up in the PET scan. This area of the brain helps us to distinguish ourselves from others. These data indicate that we come to understand another individual's behavior by imagining him or her performing the behavior and then mentally projecting ourselves into the same situation.

More comparative data are needed to determine if mirror neurons (or functional equivalents) are found in other species and if they might actually play a role in the sharing of intentions or feelings—perhaps as keys to empathy—between individuals engaged in an ongoing social interaction such as play.

Social play, social morality, and social carnivores

To stimulate further comparative research (and the development of models) on a wider array of species than has previously been studied, I offer the hypothesis that social morality, in this case behaving fairly, is an adaptation that is shared by many mammals, not only by primates. Behaving fairly evolved because it helped young animals acquire social (and other) skills needed as they mature into adults.

Group-living animals may provide many insights into animal morality. In many social groups, individuals develop and maintain tight social bonds that help to regulate social behavior. Individuals coordinate their behavior—some mate, some hunt, some defend resources, some accept subordinate status—to achieve common goals and to maintain social stability. Consider pack-living wolves. For a long time researchers thought pack size was regulated by available food resources. Wolves typically feed on such prey as elk and moose, each of which is larger than an individual wolf. Hunting such large ungulates successfully takes more than one wolf, so it made sense to postulate that wolf packs evolved because of the size of wolves' prey. Defending food might also be associated with pack-living. However, long-term research by David Mech on wolves living on Isle Royale, off the Upper Peninsula of Michigan, showed that pack size in

This photo was taken in summer of 1986 in the High Arctic of Canada's Northwest Territories. The breeding female "Mom" (lower right) has just returned and greeted the other pack members. In the photo she is trooping alongside her mate as they head from a resting area back to the den.

(Photograph courtesy of L. David Mech)

wolves was regulated by *social*, not food-related, factors. Mech discovered that the number of wolves who could live together in a coordinated pack was governed by the number of wolves with whom individuals could closely bond ("social attraction factor") balanced against the number of individuals from whom an individual could tolerate competition ("social competition factor"). Codes of conduct, and consequently packs, broke down when there were too many wolves. Whether or not the dissolution of packs was due to individuals not behaving fairly is unknown, but this would be a valuable topic for future research in wolves and other social animals.

In social groups, individuals often learn what they can and cannot do, and the group's integrity depends upon individuals agreeing that certain rules regulate their behavior. At any given moment individuals know their place or role and that of other group members. As a result of lessons in social cognition and empathy that are offered in social play, individuals learn what is "right" or "wrong"—what is acceptable to others—the result of which is the development and maintenance of a social group that operates efficiently. The absence of social structure and boundaries can produce gaps in morality that lead to the dissolution of a group.

Social play is a useful behavior pattern on which to concentrate in order to learn more about the evolution of fairness and social morality. (While birds and individuals of other species engage in social play, there are too few data from which to draw detailed conclusions about the nature of their play.) There is strong selection for playing fairly because most if not all individuals benefit from adopting this behavioral strategy (and group stability may be also be fostered). Numerous mechanisms (play invitation signals, variations in the sequencing of actions performed during play when compared to other contexts, self-handicapping, role-reversing) have evolved to facilitate the initiation and maintenance of social play in numerous mammals—to keep others engaged—so that agreeing to play fairly and the resulting benefits of doing so can be readily achieved.

Mark Ridley points out that humans seem to be inordinately upset about unfairness, but we do not know much about others animals' reaction to unfairness. Ridley suggests that perhaps behaving fairly pays off in the long run. In his book *Wild Minds*, Marc Hauser concluded that there is no evidence that animals can evaluate whether an act of reciprocation is fair. However, he did not consider social play in his discussion of animal morality and moral agency. Frans de Waal remains skeptical about the widespread taxonomic distribution of cognitive empathy after briefly considering social play, but he remains open to the possibility that cognitive empathy might be found in animals other than the great apes. It is premature to dismiss the possibility that social play plays some role in the evolution of fairness and social morality or that animals other than primates are unable intentionally to choose to behave fairly because they lack the necessary cognitive skills or emotional capacities. We really have very little information that bears on these questions.

The mystery of the mental

Lori Gruen, a philosopher at Wesleyan University, is of the opinion that we need to revisit some basic questions and come to terms with what it means to be moral. Gruen also suggests that we need to find out what cognitive and emotional capacities operate when humans perform various moral actions, then study animals to determine if they share these capacities or some variation of them. Even if it were the case that data suggested that nonhuman primates do not seem to behave in a specific way—for example, playing fairly—in the absence of comparative data this does not justify the claim that individuals of other species cannot play fairly. At a meeting in Chicago in August 2000 dealing with social organization and social complexity, it was hinted to me that while my ideas about social morality are interesting, there really is no way that social carnivores could be said to be so decent—to behave (play) fairly—because it was unlikely that even nonhuman primates were this virtuous. Perhaps it will turn out that the best explanation for existing data is that some individuals do indeed on some occasions intentionally, dare I say consciously, modify their behavior to play fairly, perhaps because of cognitive empathy.

Play may be a unique category of behavior in that inequalities are tolerated more than in other social contexts. Play cannot occur if the individuals choose not to engage in the activity, and the equality (or symmetry) needed for play to continue makes it different from other forms of seemingly cooperative behavior (e.g., hunting, caregiving). This sort of egalitarianism is thought to be a precondition for the evolution of social morality in humans. From where did it arise? Truth be told, we really do not know much about the origins of egalitarianism. Armchair discussions, while important, will do little in comparison to our having direct experiences with other animals.

The realm of the mental remains a mystery. Studying the evolution of cooperation, fairness, trust, and social morality goes well beyond traditional science and can be linked to religion, theology, spirituality, and perhaps even different notions of God because ideas about continuity and discontinuity (the possible uniqueness of humans and other species) have to be taken into account. Studying and learning about animal play can also teach us to live more compassionately with heart and love. Studies of the evolution of social morality are among the most exciting and challenging projects that behavioral scientists (ethologists, geneticists, evolutionary biologists, neurobiologists, psychologists, anthropologists) and religious scholars face. We need to rise to the task before us rather than dismiss summarily and unfairly the moral lives of other animals. Fair is fair.

seven

◳

ANIMAL WELFARE, ANIMAL RIGHTS, AND ANIMAL PROTECTION

Minding individuals with respect, compassion, and love

Imagine the following "real world" scenario. You are sitting on a dock and a young baby falls into the water. Would you jump in the water to save her? Would it make a difference if she were your sister or a stranger? Would you jump in the water to save a chimpanzee, a dog, a cat, a mouse, a pigeon, a turtle, a fly? And why do you make the choices you have? Now picture this situation. You are told that your mother will likely survive an illness if ten chimpanzees are humanely sacrificed on her behalf. What would you do? Would it make a difference if only one chimpanzee was sacrificed or if the animal who had to be sacrificed was a dog, a mouse, or a goldfish? Would it matter if the research was not painful to the animals involved and there was little or no suffering? How about if the person to be saved was a friend's mother or a total stranger?

Our beliefs about the nature of animals influence how we treat them. How we represent animals, the conditions in which we study them, and how smart or emotional we assume them to be all influence what we consider to be permissible and nonpermissible treatment. It seems that it is the complex emotional lives of animals, rather than how smart they appear to be, that more strongly influences how people choose to treat them.

In this chapter I discuss many of the issues with which we need to be concerned if we are to make this a better world in which all beings can live in harmony, a world in which respect, deep heartfelt compassion, and love replace cruelty, disrespect, and dismissal of the lives of other animals. It is impossible to write a book about minding animals without discussing the well-being of *individuals* as they strive to live and compete in a human-dominated world. I am an optimist, and while I fear that silent springs— and summers, winters, and falls—may be in the offing if we do not redefine our relationships with animals and other nature, I have enough faith in the human spirit to believe that this will not happen. I maintain unflagging hope that collectively we can make this a better world because we are a very special species, but not a *better* species than others.

Hunting and fishing: Family or spiritual experiences?

I am not going to discuss in detail such contentious issues as hunting, fishing (or dog-fighting, cock-fighting, or bullfighting). I am very sensitive to arguments about cultural traditions about how food is procured. Nonetheless, I am against all forms of sport and trophy hunting and fishing when there are alternative ways to find suitable food. In his book *The Island Within*, Richard Nelson describes what he went through when killing a deer with a rifle.

> I raise the rifle back to my shoulder, follow the movement of the deer's fleeting form, and wait until he stops to stare back. . . . I carefully align the sights and let go the sudden power. . . . The gift of the deer falls like a feather in the snow. And the rifle's sound has rolled off through the timber before I hear it. . . . I whisper thanks to the animal, hoping I might be worthy of it. . . . Incompatible emotions clash inside of me—elation and remorse, excitement and sorrow, gratitude and shame.

Nelson then guts the animal. Using beautiful prose (or poetry) to sanction the unnecessary killing of another animal being only brings me more disappointment about the ends to which some people will go to justify the spiritual uplifting that accompanies and follows the act. What about the deep suffering of the deer while being pursued and then killed? Patrick Bateson and his colleagues at the University of Cambridge in England have found that red deer stalked by dogs showed by their stress responses that they were anxious and scared. Stalked deer suffered from high levels of cortisol and the breakdown of red blood cells, indicating extreme physiological and psychological stress. Stalked deer also showed excessive fatigue and damaged muscles. Deer who were not stalked and those shot without prolonged stalking did not show similar stress responses. Clearly, animals do not like the distress, anxiety, and fear of being stalked. Neither do humans. There is nothing reciprocal in these sorts of interactions, and an increase in the spiritual life of a human who kills another animal unnecessarily brings along with it the end of any spirit or life spark that the hapless victim might have possessed.

I am baffled how taking a life unnecessarily can be anything but profoundly sad. Some people feel that hunting provides quality family bonding time. In their detailed tome about hunting and fishing, Mark Duda and his colleagues claim, "To a large extent, hunting represents the embodiment of family values." They and others also note that here is an emotional feeling, indicated by elevated heart rate and often reported as erotic or exhilarating, that is unique to the human predator-prey encounter. This excitement might also serve as a strong motivator in that stalking and hunting can be arousing and enjoyable.

Surely there are better ways to spend family time than watching animals in cages or scaring, harassing, maiming, or killing them. Children might well think about other activities in which they and their family and friends can partake, activities that are friendly to, and truly help, animals and the planet. A recent report by Jody Enck and his colleagues shows that hunter recruitment and retention in the United States is decreasing. I would like to believe that it is because people are coming to realize that unnecessarily killing other animals is wrong, it is ethically indefensible. Robert Holsman discovered that "a review of the research literature and several recent case-study examples suggests that hunters often hold attitudes and engage in behaviors that are not supportive of broad-based, ecological objectives."

Moving toward patient and compassionate activism

As an academic I often get sidetracked into intellectual arguments that *seem* to lead nowhere in terms of making the world a better place for animals. Academic musings are important, and I can get very stimulated by engaging in these exchanges. But I am also an activist, and on occasion I have to put aside all the "academic BS" (as some of my friends call it) and get out into the real world and *do* something. This was among the reasons that Jane Goodall and I founded Ethologists for the Ethical Treatment of Animals/Citizens for Responsible Animal Behavior Studies (www.ethologicalethics.org). The worldwide response to our group continues to be very positive, and we are pleased that not only professors and other researchers have joined but also lawyers, veterinarians, secondary and primary school teachers and students, and people who oversee museum and zoo exhibits.

When people discuss animal welfare and animal rights, many difficult issues come to the table. Individuals' emotions rage, and experts and those who love animals often cannot agree. It is especially upsetting to see infighting in the animal protection movement that undermines the common goal of relieving animal pain and suffering. However, what animal advocates do agree on is that there is a need to continue to encourage more compassion and love for animals and also pursue vigorously the development and implementation of nonanimal alternatives to animal use. At a recent conference in Bologna, Italy, and in a book that stemmed from it edited by Michael Balls and his colleagues (*Progress in the Reduction, Refinement, and Replacement of Animal Experimentation*), over 1,500 pages were devoted to discussion of animal use and ways to develop nonanimal alternatives. Many scientists took part in the discussions and fully suppported the goals of the meeting.

Ethical enrichment

It is in the best traditions of science to ask questions about ethics. The more difficult the questions, the better the science. Ethics can enrich our views of other animals in their own worlds and in our different worlds and help us to see that variations among animals are worthy of respect, admiration, and appreciation. The study of ethics can also broaden the range of possible ways in which we interact with other animals without ruining their lives. The separation of "us" from "them" presents a false dichotomy, the result of which is a distancing that erodes rather than enriches the possible relationships that can develop among all animal life. Ethical discussion can help us to see alternatives to past actions that have shown disrespect to other animals and, in the end, have not served us or other animals well. In this way, the study of ethics is enriching to other animals and to ourselves.

If we think that ethical considerations are stifling and create unnecessary hurdles over which we must jump in order to get done what we want to get done, then we will lose rich opportunities to learn more about other animals and also ourselves. *Our greatest discoveries come when our ethical relationship with other animals is respectful and not exploitive.*

Moderation and consistency

Most people take a moderate position on animal use by humans. They allow some but not all animal use. They feel all right about the use of some animals rather than others. For these people, all animals are not equal. They often find it difficult to be consistent and objective. Maybe it would be all right to use chimpanzees to save their own mother's or child's life, but not the life of someone else's mother or child. Perhaps it is fine to have a fish in an aquarium or a bird in a cage, but not a gorilla in a zoo.

Lisa Mighetto emphasizes in her book *Wild Animals and American Environmental Ethics*, "Those who complain of the 'inconsistencies' of animal lovers understand neither the complexity of attitudes nor how rapidly they have developed." Even with our inconsistencies and contradictions, when dealing with the difficult issues centering on animal protection, we have come a long way in dealing with many, but not all, of the problems. But we should not be complacent, for there still are far too many animals suffering at the hands of humans, and much work still needs to be done. Consistency and strong guidelines are needed to guide us so that we can lessen the pain and suffering that humans cause to other animals. Our wanton destructive tendencies do not end with animals. Our treatment of animals has large and often irreversible impacts on a wide variety of habitats, ranging from terrestrial to aquatic landscapes.

Planetary biodiversity and the loss of critical habitat: There are too many of us

Populations of humans are growing rapidly, and many populations of wild animals and plants continue losing their battle with humans. Global biodiversity—the number of different species that inhabit our planet—is rapidly, and perhaps irreversibly, dwindling. Many researchers think that the main problem is fairly simple: there are too many people and not enough land for them. Indeed, the loss of critical habitat is considered by most conservation biologists to be the biggest threat to animal and plant life. Uncontrolled habitat loss means there will be a reduction in global biodiversity. Even if humans want to reintroduce species to the wild or relocate them to suitable habitats where they would be able to thrive and survive, such places will not be available, because while the animals are not present humans continue to develop the area and make it impossible to place them there at a later time. There are too many people. There are too many people. This mantra is worth chanting over and over again. When human populations show explosive growth, it is other animals, entire ecosystems, species, populations, and individuals, that suffer.

Some difficult questions about animal use

One of the first questions that needs to be considered is "How *should* humans treat other animals?" Do we *have* to treat animals in certain ways? Are there right and wrong ways for humans to treat other animals? Can we do whatever we want to other animals? Do we need to respect animals' *rights*? Do animals even have rights? If animals have rights, what does this mean?

For many of the questions about how animals should be treated by humans, there are not "right" and "wrong" answers. However, there are "better" and "worse" answers. Open discussion of all sides will help us make progress. No one view can be dismissed by pretending that it does not exist. Ignoring the problems will not make them disappear. How we relate to animals is closely related to how we relate to ourselves and to other humans.

It is also important to remember that when humans choose to use animals, the animals invariably have no say in these decisions. They cannot give their consent. Animals depend on our goodwill and mercy. They depend on humans to have their best interests in mind.

"Them" and "us": Language and tools

For many years, people argued that it was the use of language that separated humans from other animals. But when it was discovered that there are other animals who use language to communicate with one another, language was no longer a reliable behavior to separate humans from other animals. Of course, other animals do not use human languages to communicate, but many animals use their own complex language to tell others what kinds of food are around, where they are traveling, how they are feeling, or what they need.

There was also a time when humans thought that only humans made and used tools, so this ability was used to separate humans from other animals. But when Jane Goodall, studying chimpanzees at the Gombe Stream in Tanzania, observed and filmed tool manufacture and use by David Graybeard and other chimpanzees, tool use was no longer unique to humans. Now it is known that many animals use tools.

Are humans unique? Yes, but so are other animals. The important question is *"What differences make a difference?"* What differences between individuals mean that it is all right to use or exploit one animal rather than another? If all life is respected, then it is hard to draw the line between those individuals who can be used, harmed, or killed and those who cannot. But from the practical point of view, especially if you agree that *sometimes* it is all right to use animals, then sometimes you have to make difficult decisions.

Humans are constantly making decisions for numerous animals. *We are their voices.* So when we speak for them, in order for there to be balance, we need to be sure that we are taking into account their best interests. Jane Goodall laments: "The least I can do is to speak out for the hundreds of chimpanzees who, right now, sit hunched, miserable and without hope, staring out with dead eyes from the metal prisons. They cannot speak for themselves." I cannot ask my companion dog, Jethro, if it is all right to feed him now, if it is all right to walk him now, or if it is fine for me to use him in an experiment in which he will feel pain and suffer, and perhaps be killed. When humans use chimpanzees for behavioral or medical research, they do not ask them if they agree to be kept in a small cage alone, be injected with a virus, have blood drawn, and then perhaps be "sacrificed"—killed—so that the psychological and physiological effects of the experiment can be studied in more detail.

Because of our position in the world, because we can freely speak and express our feelings about animals and animals do not have much say in the matter, animals seem to be there for us to use in any way we choose. Unfortunately, many people are largely detached from nature and the outdoors in general. A recent survey showed that many people spend more than 95 percent of their lives indoors. One result of this detachment from nature and our animal kin is that animals come to be treated as if they are property, like bicycles or backpacks, for humans to use however they choose. Animals should not be viewed as property, resources, or disposable machines that are here for human consumption. That is not why animals have evolved to be the splendid beings they are.

Some guiding principles: What is necessary?

No one *has* to do something just because he *can* do it. Just because certain activities *seem* to have worked in the past does not mean that they truly have worked. For example, many people believed that invasive experimental research on animals and eating meat were *essential* for human betterment, whereas we now know that this is not so. While some people have to eat some animal protein, even minimal amounts, most people do not. While we cannot undo all the mistakes people have made, there is still time—but perhaps not a great deal of time—to make changes that will help us and other animals along.

I consider myself lucky and privileged to have been able to meet many diverse and interesting animals. I am sure that in some instances they were watching, smelling, hearing, and studying me as closely as I was observing them. Often people are not aware that they are interfering in the lives of the animals in whom they are interested. *The guiding principles for all of our interactions with animals should stress that it is a privilege to share our lives with other animals; we should respect their interests and lives at all times, and the animals' own views of the world must be given serious consideration.*

Although I have always been concerned with the plight of animals, I have not always applied the same standards of conduct to my own research. I have done experiments that I would never do again on predatory behavior in infant coyotes in which mice and chickens were provided as bait. Coyotes were allowed to chase and kill the mice and chickens, and the mice and chickens could not escape. I am sorry I did this type of research, and I have apologized to, and said prayers for, the animals that I allowed to be killed.

I also believe that some research should not continue to be done. Included in this group are studies of learned helplessness (reactions to inescapable shock or situations in which rats are forced to swim until they are exhausted and die), social deprivation and isolation experiments, research in which animals are subjected to extreme physical restraint or starvation, staged encounters in which animals are permitted to harm or kill one another, and studies in which animals are castrated. While reasonable people can disagree about some particular lines of research, reasonable people cannot disagree about the necessity for reforming our practices with respect to animals and the changes of outlook that may be required to do so.

To make some of my points more succinctly with no further ado, I offer these two descriptions of different research projects. The first involves dogs used in studies of learned helplessness (which, on some of the authors' admission, have not been very helpful for learning about human depression, one of the reasons for which they were conducted), and the second, monkeys used in radiation research. You be the judge.

> When a normal, naive dog receives escape/avoidance training in a shuttlebox, the following behavior typically occurs: At the onset of electric shock the dog runs frantically about, defecating, urinating, and howling until it scrambles over the barrier and so escapes from the shock. . . . However, in dramatic contrast . . . a dog who had received inescapable shock while strapped in a Pavlovian harness soon stops running and remains silent until shock terminates. . . . It seems to "give up" and passively "accept the shock."

> In one set of tests, the animals had been subjected to lethal doses of radiation and then forced by electric shock to run on a treadmill until they collapsed. Before dying, the unanesthetized monkeys suffered the predictable effects of excessive radiation, including vomiting and diarrhoea. After acknowledging all this, a DNA [Defense Nuclear Agency] spokesman commented: *"To the best of our knowledge, the animals experience no pain."*

Some questions about animal use

Here are some questions to consider about animal use. Many of the questions are related to one another, and it is difficult to discuss one without considering others.

Why might some people think that is all right to kill dogs rather than
other people?

Why do some people feel more comfortable killing ants rather than
dogs?

Are some species more valuable or more important than others?

Do some animals feel pain, experience anxiety, and suffer, and not
others?

Are some animals conscious?

Do animals feel emotions?

Are some animals smart and not others?

Do some animals have rights and not others?

What is the difference between animal rights and animal welfare?

Should endangered animals such as wolves be reintroduced to places
where they originally lived?

Should we be more concerned with species and their survival than
with individuals and their well-being?

Should animals be kept in captivity, in zoos, wildlife theme parks, and
aquariums?

Should we interfere in the lives of other animals?

Should we interfere in fights in which an individual could get
hurt? Feed starving animals? Give first aid when animals are hurt?
Rescue animals from oil spills? Inoculate animals to protect them
from diseases such as rabies?

Why do some humans eat animals?

Why do some humans use animals for research?

Why do many people feel more comfortable using dogs rather than
other people for research?

Should animals be used to test cosmetics or foods?

Should domesticated animals such as dogs and cats be treated differently than their wild relatives, wolves and lions?

Do we need to cut up animals—dissect them—to learn about them
or ourselves?

What types of nonanimal alternatives are available for product testing, dissection, and vivisection?

Where do we go from here?

Human and nonhuman primates: How close we are

If we conclude that chimpanzees are conscious, we must then confront
the ethics of our treatment of such animals in captivity and in the remaining wild.

—Alison Jolly, "Conscious Chimpanzees?"

Although there are numerous differences between humans and other animals, in many important ways "we" (humans) are very much one of "them" (animals), and "they" are very much one of "us." For example, researchers have compared proteins on the surface of human and chimpanzee cells. Of nine amino acid chains (amino acids are the building blocks of proteins) studied, there are only five (0.4 percent) differences out of a total of 1,271 amino acid positions. This means we are 99.6 percent chimpanzee and vice versa. (The word "chimpanzee" means "mock man" in a Congolese dialect.) Also, humans and chimpanzees share 98.7 percent of their genes. Gorillas are 2.3 percent different from both humans and chimpanzees, and orangutans are 3.6 percent different from both humans and chimpanzees.

Despite the closeness between humans and our next of kin, the great apes, commonly, when humans and these and other animals' paths cross, the animals lose. Chimpanzees are used in much research that causes pain and suffering. They are strapped to chairs and unable to move, subjected to radiation that makes them violently ill and sometimes kills them, shocked in situations from which they are not allowed to escape, and injected with infections that sicken and kill them. They are also used to study diseases, such as AIDS, from which they do not suffer. When they are used in these sorts of experiments, they are often housed alone in very small cages and suffer severe emotional stress (depression) as well as physical trauma (great weight loss, self-mutilation). Worldwide there are movements, such as the Great Ape Project, to make their lives better and to eliminate their experimental use altogether. The British and New Zealand governments have decided that great apes can no longer be used in research. France and other European countries have signed the Treaty of Amsterdam, which recognizes many animals as sentient beings capable of feeling fear and pain and of enjoying themselves. Even octopuses, with their large central nervous system and complex behavior, have been given the benefit of the doubt in Great Britain. Since 1993 they have been protected under the Animals (Scientific Procedures) Act of 1986 that regulates the use of animals in scientific research. There are also efforts to build retirement homes for chimpanzees and other animals when their research careers are over so that the animals have the best lives possible and die a natural death.

The Great Ape Project: Granting apes legal rights

In 1993, a book titled *The Great Ape Project: Equality Beyond Humanity* was published. This important book launched what has become known as the Great Ape Project (GAP). The major goals of the GAP were to admit great apes to the *Community of Equals* in which the following basic moral rights, enforceable by law, are granted: (1) the right to life, (2) the protection of individual liberty, and (3) the prohibition of torture. In the GAP, "equals" does not mean any specific actual likeness but equal moral consideration.

Some people think that the GAP does not go far enough, because of its speciesist concentration on great apes to the exclusion of other animals. But beginning with animals who would generate the least resistance was probably the correct choice. While many people might be willing to grant certain legal rights to great apes, many would not want to grant them to dogs, cats, birds, mice or other rodents (many of whom are used in research), fish, crocodiles, lobsters, or ants.

Do some animals experience pain, anxiety, and suffering?

I hesitate to ask this question, for the answer is so evident it seems like one of those academic exercises in which people engage because they have nothing better to do. Pain is an unpleasant sensation or range of unpleasant sensations that can protect animals from physical damage or threats of damage. For example, when animals are bitten hard they move away from the animal biting them. To experience pain an individual must have at least a simple nervous system. There is no doubt that many animals experience pain. Veterinarians have developed a pain patch for dogs coming out of surgery. If they did not know that the dogs felt pain and suffered, why would they have developed this patch? We have all heard dogs yelp when they step on a nail, catch their tail in a door, or are bitten too hard. The University of Tennessee College of Veterinary Medicine has established the Center for the Management of Animal Pain to improve methods of preventing and treating pain in animals. Even if animals (including humans) do not know who they are, this does not mean that they do not experience pain. I agree with Georgia Mason, who points out that there seems to be no good reason why self-awareness needs to be a prerequisite for suffering, why "the (self-aware) feeling 'I am suffering' [should] be considered worse than the (not self-aware) 'Something truly terrible is happening.' "

The experience of pain is unavoidable. Pain serves many useful functions and contributes to survival. Humans and animals lacking pain systems by accident of birth or disease tend to have shorter lives. While researchers are not sure which animals feel pain there is much evidence that animals who many people thought could not feel pain in fact, do. Fish, for example, have neurons similar to those that are associated with the perception of pain in other animals. Fish show responses to painful stimuli that resemble those of other animals, including humans. Even some invertebrates possess nerve cells that are associated with the feeling of pain in vertebrates. Whether some insects actually feel pain is not known, but because they might some people believe that they should be given the benefit of the doubt.

Animal welfare and animal rights

> If we, in the western world, see a peasant beating an emaciated old donkey, forcing it to pull an oversize load, almost beyond its strength, we are shocked and outraged. But, taking an infant chimpanzee from his mother's arms, locking him into the bleak world of the laboratory, injecting him with human diseases—this, if done in the name of Science, is not regarded as cruelty. Yet in the final analysis, both donkey and chimpanzee are being exploited and misused for the benefit of humans. Why is one any more cruel than the other? Only because science has come to be venerated, and because scientists are assumed to be acting for the good of mankind, while the peasant is selfishly punishing a poor animal for his own gain.
>
> —Jane Goodall, *Through a Window*

What does it mean if animals can feel pain? If animals feel pain and are able to suffer, then we must be careful not to cause them unnecessary pain and suffering. While some people believe that it is all right to cause animals pain if the research helps humans, there are others who believe that this should not be done even if humans might benefit from the research.

People who believe that we are allowed to cause animals pain, but that we must be careful not to cause them excessive or unnecessary pain, argue that if we consider the animals' *welfare* or *well-being*, then that is all we need to do. These people are called *welfarists*. Those people who believe that it is wrong to cause animals any pain and suffering, and that animals should not be eaten, held captive in zoos, or used in painful research, or in most or any research, are called *rightists*. They believe that animals have certain moral and legal rights that include the right not to be harmed.

Many people support a position called the *rights* view. According to the lawyer and animal rights advocate Gary Francione, to say that an animal has a "right" to have an interest protected means that the animal has a claim, or entitlement, to have that interest protected even if it would benefit us to do otherwise. Humans have an obligation to honor that claim for other voiceless animals just as they do for young children and the mentally disabled. So if a dog has a right to be fed, you have an obligation to make sure she is fed. If a dog has a right to be fed, then you are obligated not to do anything to interfere with feeding her.

Tom Regan, a professor of philosophy at North Carolina State University, is often called the "modern father of animal rights." His book *The Case for Animal Rights*, published in 1983, attracted much attention to this area. Advocates who believe that animals have rights stress that animals' lives are valuable in and of themselves, not because of what they can do for humans or because they look or behave like us. Animals are not property or "things" but rather living organisms, subjects of a life, who are worthy of our compassion, respect, friendship, and support. Rightists expand the list of animals to whom we grant certain rights. Thus animals are not "lesser" or "less valuable" than humans. They are not property that can be abused or dominated.

Many people think that the animal rights view and the animal welfare view are the same. They are not. Most welfarists do not think that animals have rights. Some welfarists do not think that humans have rights either. Rather, they believe that while humans should not abuse or exploit animals, as long as we make the animals' lives comfortable, physically and psychologically, then we are taking care of them and respecting their welfare. Welfarists are concerned with the quality of animals' lives. But welfarists do not believe that animals' lives are valuable in and of themselves, just because animals are alive.

Welfarists believe that if animals experience comfort, appear happy, experience some of life's pleasures, and are free from prolonged or intense pain, fear, hunger, and other unpleasant states, then we are fulfilling our obligations to them. If individuals show normal growth and reproduction and are free from disease, injury, malnutrition, and other types of suffering, they are doing well.

This welfarist position also assumes that it is all right to use animals to meet human ends as long as certain safeguards are employed. They believe that the use of animals in experiments and the slaughtering of animals for human consumption are all right as long as these activities are conducted in a humane way. Welfarists do not want animals to suffer from any unnecessary pain, but they sometimes disagree among themselves about what pain is necessary and what humane care really amounts to. But welfarists agree that the pain and death animals experience are sometimes justified because of the benefits that humans derive. The ends (human benefits) justify the means (the use of animals even if they suffer) because the use is considered to be necessary for human benefits.

Balancing the costs and benefits of animal use

In the real world in which practical decisions constantly have to be made, many people practice welfarism. They think of animals with respect to how they may serve humans, how they may be used in research, for food, or for amusement or entertainment. Because this view of animals is so prevalent, it is worth seeing how welfarism works.

People who consider animals' usefulness to humans are called *utilitarians*, and they practice *utilitarianism*. The philosopher Peter Singer, author of *Animal Liberation*, who now teaches at Princeton University, is the modern-day champion of utilitarianism as it relates to how humans use other animals. Utilitarians believe that a dog, cat, or any other animal can be used as long as the pain and suffering that the animal experiences—the *cost* of using the animal to the animal—is *less* than the *benefits* to humans that are gained by using the animal. Singer believes that the best course of action is the one that has the best consequences, on balance, for the interests of all those who are affected by a particular decision to do something or not to do something. He stresses that the interests of animals must be given equal consideration with those of humans and that both animals and humans have an interest in avoiding suffering.

When utilitarianism is applied to animals, it is much the same as *welfarism*. The only rule, and it is not a moral rule, is that it is all right to use animals if the relationship between the costs to the animals and the benefits to the humans is such that the costs are less than the benefits. Utilitarians may argue that it is all right to use a million mice in cancer research to save only one human life because the costs to the mice are less than the benefits to the human(s) who might use a treatment that was developed using the mice. Or they may believe that it is all right to keep gorillas or other animals in cages in zoos because the costs to the animals are less than the benefits to the humans who supposedly learn about the animals' lives.

One major problem with utilitarianism concerns how costs and benefits are calculated. How does one decide that the pain, suffering, and lives of a million mice cost less to the mice than the benefits that are gotten by one or more humans? Why not balance a million mice with a hundred humans, or a million mice with a hundred thousand humans? Because it is humans who are making the decisions about costs and benefits, there is always the chance that there will be some bias in favor of humans. One reason many people are not satisfied with utilitarianism is just that: because it is humans who make the decisions, it is easy to make the equation always come out in favor of the humans. An animal's interest can be ignored if it benefits us to do so.

According to the standard version of utilitarianism, first offered by the English philosopher Jeremy Bentham, what really matters is pleasure or pain. Bentham was very interested in animals and wanted animals to be included in moral decisions made by humans. Because of his concern with animals, he wrote: "The question is not, Can they *reason*? nor Can they *talk*? but, Can they *suffer*?" For Bentham, it really did not much matter if animals could think or if they were smart. Rather, Bentham was concerned with whether or not animals could suffer. It is the costs associated with suffering that need to be considered when deciding how costs and benefits are balanced. Utilitarians who follow Bentham's ideas judge an action as right if it leads to greater pleasure than pain. People should aim to maximize pleasure, the benefit, and minimize pain, the cost.

Here are some of the variables that play a role in using utilitarianism in making decisions about animal use: (1) who the person is who makes the decision; (2) who the people are who might benefit; (3) which animals are used; and (4) how the animals are to be used. One factor that needs to be included, but one that often is not, concerns possible benefits to the animals who are used or benefits to other members of the same or other species. Perhaps the individual mice or chimpanzees who are used will benefit from the research if they survive the procedures and do not suffer pain and injuries from which they cannot recover. Perhaps other mice and other chimpanzees will benefit from the pains, suffering, and death of other mice and chimpanzees. So it is for the good of the species that some individuals suffer or are killed.

Zoos, theme parks, and aquariums

Zoos have existed for a long time. Ancient Egyptians are known to have kept collections of animals. The first modern zoos were in Europe and opened in the late 1700s. In the United States, the first European-style zoo opened in 1874 in Philadelphia. It was modeled after the London Zoo. The first public aquarium in the United States was opened in 1856 by P. T. Barnum, of circus fame. Early zoos were primarily for the humans and not the imprisoned animals. They were essentially living museums.

In the United States, the American Zoo and Aquarium Association (AZA), incorporated in 1972, is responsible for inspecting zoos, wildlife theme parks, and aquariums (collectively referred to as zoos). If these institutions meet AZA standards, they are approved and accredited by the AZA. There are fewer than two hundred accredited zoos in the United States and about two thousand licensed zoos that are not accredited by the AZA.

Zoos vary greatly in quality. Some zoo experts feel that many zoo exhibits are antiquated and that only about one-third could generously be called "enriched" or "naturalistic." One zoo director says that he would like to be able to change 95 percent of the exhibits he has visited. A 1995 Roper poll showed that about 70 percent of Americans were concerned about the well-being of animals in zoos. Many studies show that captive animals are generally stressed, but this varies among different species and how they are housed. While there are some zoos that are trying as hard as they can to make the lives of their residents the best they can be, there are numerous zoos that are of very poor quality. When making decisions about whether or not zoos should exist, we must take into account not only that *all* captive animals have compromised lives and are being kept without their permission but also that there are some zoos where the animals simply are treated poorly.

"Surplus" animals

In addition to the fact that captive animals lead unnatural lives that often are impoverished, some zoos sell, trade, donate, or loan unwanted, or "surplus," animals to animal dealers, auctions, hunting ranches, unidentified individuals, unaccredited zoos, and game farms whose owners actively deal in the animal marketplace. Most zoo animals are like museum specimens in that they will never be freed from captivity. The animals are treated as mere property, their fate dependent on their dollar value as if they were a pair of extra shoes. From 1992 to the middle of 1998, about a thousand exotic animals were sold as live merchandise. There is a vast underground involved in trafficking in rare and endangered species.

Naturalistic exhibits and environmental enrichment

Many zoos have what they call "naturalistic" exhibits, and wildlife parks provide animals with areas in which to roam that resemble their natural habitats. There is an attempt to meet the physical and behavioral needs of the caged animals by providing them with an environment that as closely as possible resembles their natural environment, from which they may have been taken or in which wild relatives still live. Wolves often are kept in groups that resemble wild packs, and animals who are more solitary are allowed to live alone and have places into which to escape or hide if they want to get away from other animals or from human spectators. However, while American brown bears in captivity spend about the same percentage of time as wild bears being active, the bulk of captive bears' activity is pacing, not foraging.

Some zoos attempt to provide captive animals with enriched environments that stimulate and challenge the animals and reduce the boredom of being in the same place with little to do day in and day out. Enrichment programs give animals control over their environment and often provide them with choices of different activities in which to engage.

Many different types of enrichment have been used. Different species have different needs, and there are individual differences within species (for example, age and gender may influence what works). Enriching animals' lives can be accomplished in numerous ways including providing safe and secure places for resting, sleeping, and escaping from unwanted intrusions by cage mates and humans, allowing individuals of social species to live in pairs or larger groups that resemble natural groups, making them work for meals by providing frozen food or by scattering or hiding it, providing natural substrates, spraying various odors in cages, allowing or making it easy to exercise, providing large cages (although this is not always enriching), and increasing the complexity or diversity of their social and or physical environments. Moderate levels of stress that tax the animals might be beneficial, especially for animals who are to be reintroduced to the wild where life is stressful. It might also be helpful to make individuals work hard for food if food is going to be difficult to obtain in the wild, or expose them to predator-like situations so that they learn to avoid predators.

While enrichment programs do not directly address the question of whether or not humans should hold other animals captive, the captive individuals who have the opportunity to experience enriched environments seem to be happier or more content than those who do not. Animals who are in good psychological health engage in less stereotyped pacing, cage-rage and self-mutilation, and rocking back and forth and show less fear and aggression than unhappy or discontented individuals. Content animals play, have good appetites, and do not suffer from the same abnormally high levels of stress, anxiety, or disease as their less fortunate kin.

Education, conservation, biodiversity, and endangered species

Two common reasons given to justify the existence of zoos are education and conservation. Some people believe that zoos are good because they educate people about animals in general and also about animal species that they would otherwise never get to see. However, Michael Kreger, at the Animal Welfare Information Center, found that the average visitor spends only about thirty seconds to two minutes at a typical exhibit and only reads some signs about the animals. A number of surveys have shown that the predominant reason people go to the zoo is to be entertained. In one study at Edinburgh Zoo in Scotland, only 4 percent of zoo visitors went there to be educated, and no one specifically stated he or she went to support conservation. There is very little evidence that much educational information is learned and retained that helps the animals in the future. In another venue, Alan Beck and his colleagues discovered that a home-based educational program for feeding wild birds led to an increase in knowledge about birds for seven- to nine-year-old boys and girls but not for ten- to twelve-year-olds.

Some people support zoos because they might serve to keep individuals of rare, threatened, or endangered species alive when the habitat of these animals has been destroyed. However, it has been estimated that about 50 percent to 70 percent of orphaned gorilla infants who are taken into captivity are likely to die. The figures are similar for orphaned gorilla infants and juveniles who are released from captivity into the wild.

Sometimes individuals are kept in zoos because at some time in the future either they or their offspring will be released into the wild. Thus some people think that zoos are valuable because they will help maintain biodiversity. They argue that without zoos, biodiversity will decrease as species go extinct. Thus zoos potentially can be important in conservation efforts by keeping animals in safe places and then releasing them into the wild. But if habitat is not preserved for them, if people use the land for other purposes while the animals are held in captivity, there will not be anywhere for them to be released. This happens quite often. Indeed, most conservation biologists agree that habitat loss is the major cause for losses of biodiversity. There are too many people and too little land for animals to thrive and survive. The situation is not getting better. In Kenya it is estimated that wild lands are disappearing at a rate of about 2 percent a year.

Zoos actually do little for increasing biodiversity. While some zoos make serious efforts in the conservation arena, few zoos actually have conservation programs, and in those that do, only a small percentage of the zoo's budget is spent on these programs. In a period of ten years, the San Diego Zoo reported that it spent $55 million on public relations but only $17.6 million on wildlife conservation studies.

There is little evidence that release/reintroduction programs using animals who have been former residents of zoos or their offspring are successful. While successful reintroductions have been performed for Arabian oryx and perhaps golden lion tamarins in Brazil, there have not been many other programs that seem to have made any long-term difference. Of 145 reintroductions involving 126 species and 13 million individuals born in captivity, only 16 (11 percent) have succeeded. Benjamin Beck, when he was chair of the AZA's Reintroduction Advisory Group, lamented, "We must acknowledge frankly at this point that there is not overwhelming evidence that reintroduction is successful." He also noted that we just do not know enough to have successful rehabilitation and release programs for apes in captivity.

Other people who have worked in zoos also agree that conservation is not well served by zoos. In her book *The Modern Ark: The Story of Zoos, Past, Present and Future*, Vicki Croke quotes Terry Maple, director of Zoo Atlanta, as saying, "Any zoo that sits around and tells you that the strength of zoos is the SSP [Species Survival Plan] is blowing smoke." David Hancocks, director of Australia's Open Range Zoo in Werribee, claims that "zoos can immediately stop degrading the word 'conservation' by employing it so irresponsibly." Michael Robinson, a past director of the National Zoo in Washington, D.C., agrees. Most zoos concentrate on vertebrates, and innumerable imperiled species of invertebrates cannot be saved by breeding programs in zoos.

Nature's supposed "cruelty" and interfering in the lives of wild animals

Most people believe that only humans have obligations to other animals. Animals themselves do not have obligations to other animals. Because animals do not know "right" from "wrong" or "good" from "bad," when they do something that we would call "bad"—for example, wolves killing a moose—we cannot say that they are doing something wrong. The case of a rare white Bengal tiger who killed a zookeeper at the Miami, Florida, zoo brought up important issues concerning species differences in knowing right from wrong. It was decided that the tiger would not destroyed, because in killing the zookeeper "the tiger was just being a tiger." He was not responsible for his actions; he did not know right from wrong. However, when wolves are reintroduced to an area where they once lived, they often are killed for preying on livestock, for "being wolves."

Philosophers refer to animals as "moral patients" and humans as "moral agents." Moral agents are responsible for their behavior. Infants and senile adults who are not responsible for their own lives cannot be held responsible for other humans' or animals' lives. They are not considered to be moral agents. As moral patients, animals are not held accountable for their actions. Even if some animals, some of the time, behave as if they are moral agents, it is unlikely that their agency would be as extensive or as well developed as ours. Using nature's supposed cruelty to justify the use of animals by humans is wrong.

Animals as food

One of the most common uses of animals is for food. About five million dairy cows are kept in confinement in the United States. Female dairy cows are forced to have a calf every year. Their calves are removed from them immediately after birth so they do not drink their mother's milk. This is extremely demanding on their bodies and on their psychological states. These dairy cows are literally milk machines, and they are not allowed to be mothers, to care for their young.

Animal meat also is popular. Each year millions of animals are bred, transported, and housed in slaughterhouses waiting to be killed (and often watching other animals being brutally slaughtered). An essay in the *Washington Post* published on 10 April 2001 titled "They Die Piece by Piece" concluded that the humane treatment of cattle is often a lost battle.

Poultry meat and eggs are now the most abundant and least expensive animal food products, because of the development of a large-scale industry devoted to poultry production. Birds are kept in tiny, barren battery cages and cannot perform behaviors such as dustbathing, perching, and nesting. Many birds also have their beaks trimmed to reduce injuries and mortality associated with feather-pecking and cannibalism. About half of the beak is removed, using a hot cauterizing blade or a precision trimmer. The pain associated with beak-trimming is intense and long-lasting. Caged birds often develop osteoporosis (weakened bones) because of a lack of exercise combined with calcium deficiency associated with their high rate of egg-laying. Up to about 25 percent of hens sustain broken bones when they are removed from their cages to be transported to a processing plant. Each hen now lays upwards of 300 eggs per year, as compared to 170 in 1925.

Food animals can suffer physical and emotional pain throughout their lives, often made worse by methods used to make them "better food." Individuals being fattened up for human consumption are given various hormones and kept in crowded and restricted housing conditions. Broilers now grow to market weight in about six rather than sixteen weeks. Many animals die from stress and disease before being slaughtered. Often, fully conscious chickens and turkeys are shocked, drowned in an electrified bath of water, and scalded as they are being prepared for market. Pigs and cattle are supposed to be stunned before being hung upside down by their hind legs, having their throats slit, and bleeding to death, but often they are awake during the whole process despite a federal law that requires them to be killed humanely.

There also are genetic engineering programs that produce animals that are bigger and more meaty. There is as little regard for the rights of these "bigger and better" animals as there is for their "normal" relatives. Dairy cows stimulated with bovine growth hormone (rBGR) can produce as much as a hundred pounds of milk a day, about ten times more than they would normally yield. They suffer from udder infections and are treated with antibiotics as they continue to be exploited for milk. The antibiotics can be transferred to their milk and consumed by people.

Cows, grain, and human starvation

Raising animals for food requires a lot of food to feed the animals who are to be eaten, and also a lot of land to keep the animals and to raise the grain that is used to feed them. It has been estimated that it takes eight or nine cattle a year to feed *one* average meat-eater. Each cow needs one acre of green plants, corn, or soybeans a year. Thus it takes about nine acres of plants to supply one meat-eater, rather than half an acre for a person who does not eat meat. The amount of grain that is needed to provide meat for one person is enough to feed about twenty people enough grain to live for a year.

In the United States alone, livestock eat enough grain and soybeans each year to feed over a billion people. It takes about sixteen pounds of grain to make a pound of beef. A reduction of meat consumption by only 10 percent would result in about twelve million more tons of grain for human consumption. This additional grain could feed most, if not all, of the estimated sixty million humans who starve to death each year.

Public opinion trumps veal

Veal, which comes from cattle calves who are imprisoned in cages so small that they cannot move, may be the best example of extreme animal abuse for a food that is not needed by anyone. Veal is nonessential food. Most formula-fed veal calves are raised in tiny twenty-four-inch-wide crates for their entire lives, sixteen to eighteen weeks, and fed a liquid diet twice a day. Iron intake is restricted to below normal levels, so the calves become anemic. Anemia results in a pale or white color of the meat, and it is the paleness of a carcass that is the most important factor in grading the meat and the price paid to the producer. The production and demand for formula-fed veal has dropped sharply since 1985 and has now stabilized at approximately eight hundred thousand calves per year, a decrease of over 400 percent. Public outrage over how veal calves are treated was the major reason for this decline. It is very clear that what people felt about the lives of veal calves made a difference. In a consumer-driven society, buyers are powerful.

Vegetarianism

There are many alternatives for the vast majority of people who choose to eat meat. Vegetarian diets are much healthier than diets that contain meat, especially meat that has been injected with various types of hormones or meat from animals who were stressed before they were killed. Colin Campbell, in his long-term study of dietary habits in mainland China, has shown that a low-fat (10 percent to 20 percent of total calories), plant-based diet could significantly decrease the occurrence of chronic degenerative diseases such as various cancers and heart disorders in Western countries.

Due to the animal cruelty involved in meat-eating, many people choose to reduce or eliminate their consumption of meat. There are many types of vegetarianism and numerous reasons for becoming a vegetarian. Vegetarians can be classified as follows: *lacto-ovo vegetarians*, who eat eggs and dairy products but no meat; *lacto-vegetarians*, who eat dairy products but no eggs or meat; *ovo-vegetarians*, who eat eggs but no dairy products or meat; *vegans*, who consume no meat, dairy products, or eggs; *macrobiotic vegetarians*, who live on whole grains, sea and land vegetables, beans, and miso; *natural hygienists*, who eat plant foods and combine foods in certain ways and believe in periodic fasting; *raw foodists*, who eat only uncooked nonmeat foods; and *fruitarians*, who eat fruits but also nuts, seeds, and certain vegetables.

The philosopher Michael Allen Fox lists the following arguments for vegetarianism: (1) health; (2) to reduce animal suffering and death; (3) to promote impartiality and universal well-being; (4) environmental concerns; (5) to promote universal compassion and kinship with other animals; and (6) religious arguments. He notes that vegetarianism may be seen not only as a means of focusing attention on human-animal or human-nature relationships but also as the choice of a way of life that is morally and ecologically preferable.

Vegetarianism is on the rise worldwide. Some people become vegetarians because they care about human starvation. For example, a lacto-ovo vegetarian diet would feed the world's human population more efficiently than a meat-eater diet because a cow must eat many pounds of vegetable matter to grow a pound of meat, and much of that vegetable matter could be used to feed humans. Many people become vegetarians out of concern for the well-being of farm animals. Recent catastrophes such as the foot-and-mouth disease epidemic across Europe might make more people turn toward an animal-free diet—not only because they are afraid of eating tainted meat but also because they saw both firsthand and on newscasts the pain and suffering of the animals who were being slaughtered.

Great apes as bushmeat: Logging, hunting, and making music

Bushmeat, the meat of wild animals caught and killed in their home forests, is a very popular commercial food product in many parts of the world. Chimpanzee and gorilla meat is favored, as is kangaroo meat. About 20 percent of bushmeat is primate meat. Its consumption (even in elegant European restaurants) and trade is the biggest threat to biodiversity and endangered species in some African forests. In the Congo Basin, bushmeat is the primary source of animal protein for the majority of families.

There simply are not enough chimpanzees or gorillas to sustain their slaughter for food. In one study, it was found that about eight hundred gorillas were killed each year in the Kika, Moloundou, and Mabale triangle in Cameroon. If only three thousand gorillas live in that ten-thousand-square-kilometer area, the taking of this many gorillas simply is not sustainable. About four hundred chimpanzees were also killed in this area. Thus, twelve hundred great apes in one small area were killed. Anthony Rose, at the Biosynergy Institute in Hermosa Beach, California, estimated that in the year 2000 more than three thousand gorillas and four thousand chimpanzees were illegally butchered. That is five times the number of gorillas on Rwanda's Mt. Visoke and twenty times more chimpanzees than live near Tanzania's Gombe Stream, Jane Goodall's study area. More great apes are eaten each year than are now kept in all the zoos and laboratories worldwide. Numerous young are also orphaned when their parents are killed for meat.

The bushmeat trade is an example of how different human activities that seem to be unrelated actually influence one another greatly. For example, the large increase in the availability of bushmeat is a result of increased logging activities. Logging companies build new roads into areas that previously were very difficult to reach, and they allow hunters to travel on company vehicles to hard-to-reach areas where gorillas, chimpanzees, and other large animals are found. The hunters kill all but the smallest animals and carry meat to logging camps, where loggers consume some of it. Remaining meat goes to market in cities. As logging increases, so does the number of individuals who are brutally slaughtered. There now are campaigns to have logging companies stop the transport of bushmeat and to stop logging operations. Some efforts have been successful. In July 2001, Congolaise Industrielle des Bois, a private timber company, returned land to which it had a legal right to the Republic of Congo. This unprecedented move was negotiated by government officials and conservationists including members of the Wildlife Conservation Society based in New York.

The killing of great apes is illegal in every country in which it takes place, but very few hunters are ever punished. There are international laws concerning the killing of endangered species (for example, chimpanzees), but it is difficult to catch hunters in action. Although people have known about the harmful effects of the bushmeat trade for years, until recently there has not been much interest in this activity. But now, most conservationists realize that if commercial bushmeat hunting continues there will be devastating effects on the population of chosen animals. Thus there is a lot of interest in stopping the bushmeat trade.

Logging is not only a problem in Africa. Many conservation biologists are supporting programs that limit commercial logging in numerous countries so that habitat and animals can be protected. One way to decrease the need for wood is to be careful when purchasing wood products and to ask about the source of the timber. It is possible that the wood that you are buying in some way contributed to the death of great apes and other animals who became easy to find and kill because roads were built for other reasons. The Rescued Wood Bowl Company in Fort Collins, Colorado, founded by Trent Bosch, is setting an example by using only rescued and recycled wood, wood that was on the way to a landfill. The Gibson Guitar Company uses "smartwood" instead of wood from dwindling rain forests. Who would have ever thought that playing music with a guitar may be directly related to killing animals, not to mention killing lovely trees and decimating fragile habitats?

Although many of us live far away from places where bushmeat is slaughtered and consumed, we can protest this illegal activity by being careful about what we buy and by expressing our outrage about this carnage. Each of us counts, and we all can make a difference if we act.

Animals as clothing

Wearing animals as clothing is a common practice. Wild fur-bearing animals, over forty million individuals per year, are cruelly captured, injured, and killed for profit. Many are trapped using contraptions that cause psychological and physical suffering. These devices include leg hold traps, wire snares that encircle an animal and pull tighter as the animal struggles, and conibears that grip the entire body and break the neck or back. Beavers are often trapped in water and drown after struggling for some time. Often dogs are caught in traps set to capture other species. There are no laws in the United States regarding how trapped animals can or cannot be killed.

Animals are also raised on farms (ranch-raised) only to be slaughtered for clothing. Recently dogs and cats (bred specifically for use as clothing, or strays) have been used to make fur products. These individuals typically are kept in deplorable conditions before being killed by being beaten, hanged, suffocated, or bled to death. In the United States there are no federal laws prohibiting the import of dog or cat fur or its use in clothing.

There are no laws in the United States that regulate fur farms. Needless to say, farmed animals suffer from all the same maladies as do captive animals in zoos. (In England, Georgia Mason and her colleagues discovered that mink may thrive in captivity but are frustrated when they are unable to swim.) However, farmed animals are killed and are not even able live out their lives in cages. The fur industry has a set of guidelines, but their use is voluntary, and there is no monitoring of fur farms. Animals such as mink are killed by neck-snapping ("popping"). They show great distress when removed from their cages to be killed—screeching, urinating, defecating, fighting for their lives. Gassing is also used, as are lethal injections, both of which cause pain and prolonged suffering before animals are blessed with death. The carcasses of some farmed animals are even sold for dissection, so there is a connection between raising animals for fur and their use in education. Supporting dissection can also support the fur industry.

By going to the mass media and reaching millions of concerned consumers about the treatment to which animals are subjected in the clothing industry, people have saved the lives of numerous animals. A number of clothing designers no longer manufacture fur products. Nonetheless, much more needs to be done, for fur-farming remains an industry that causes untold pain and suffering to numerous animals.

Alternatives to the use of animals

Because of increasing pressure to reduce the use of animals in research and education, many people are interested in developing nonanimal alternatives. The idea of what have come to be called the "Three *R*'s," *reduction, refinement, and replacement*, first appeared in a book published in 1959 titled *The Principles of Humane Experimental Technique*, written by two British scientists, William M. S. Russell and Rex Burch. This book was the first to present in a clear fashion how animals could be protected from human abuse. Many researchers use the Three *R*'s today.

Reduction alternatives use fewer animals to obtain the same amount of data or allow more information to be obtained from a given number of animals. The goal of reduction alternatives is to decrease the total number of animals that are used.

Refinement alternatives lessen animal pain and distress. When developing refinement alternatives it is important to assess the level of pain an animal is experiencing. It is assumed that if a procedure is painful to humans, it will also be painful to animals. Refinement alternatives include the use of analgesics and/or anesthetics to alleviate any potential pain. Environmental enrichment, discussed above, is also an example of refinement.

Replacement alternatives are methods that do not use live animals, such as in vitro—meaning "in glass"—systems. In vitro studies use living material or parts of living material cultured in petri dishes or in test tubes. ("In vivo" studies are those carried out "in living animals.") Mathematical and computer models also can be used to replace animals.

Dissection

One area in which there is much ongoing discussion of nonanimal alternatives concerns the use of animals in education. For example, opinions vary on whether dissection (cutting apart dead animals) or vivisection (experimenting on live animals) is essential to learning about animals. Individuals of about 170 species, including at least ten million vertebrates, are used annually for education in the United States. It has been estimated that about 90 percent of the animals used for dissection, including frogs, turtles, and fish, are wild-caught. The philosopher Stephen Sapontzis has pointed out that killing and dissecting can teach bad attitudes about animals and can lead students to think that animals are weak, that exploitation of the weak by the strong is all right. Teachers and students can also become desensitized to the plight of other animals and lose respect for the animals who are being used.

Many schools require students to dissect dead animal specimens. Most students do not say anything to teachers about their objections to dissection, and many students do not know that there are numerous nonanimal options that are readily available. They are not told about alternatives but rather have to ask about them. For example, all twenty-four county school systems in Maryland once permitted students to use alternatives to dissection, but only one county had a written policy that required students and/or parents to be notified of this option. Ridicule, humiliation, lost time, and perhaps feeling that one has to change one's career choice also may make students decide to do something that they do not want to do.

Supporters of dissection, including the Human Anatomy and Physiology Society, argue that "hands-on" dissection experience is essential to the student's education. Some biologists think that if someone does not want to cut animals up, he or she should not study biology. These scientists overlook the fact that there are many different types of biology that range from anatomical or physiological studies to watching how animals behave.

There is no evidence for the claim that hands-on dissection is essential to the student's education. Many teachers who use dissection in their classrooms do so despite the fact that they do not know if exercises such as these actually work. Appeals to history, saying that this is the way we always did it, often provide weak reasons for continuing practices that either should never have been freely implemented in the first place or are simply outdated because of advances in other fields.

Many medical schools have come to recognize that hands-on experience is not needed in certain parts of their curricula. Currently, 90 of 126 (71 percent) American medical schools, including such prestigious institutions as Harvard, Yale, Columbia, Duke, and Stanford, do not use live animal laboratories in training medical students. Medical students can select nonanimal options or choose not to work in a particular laboratory that uses animals even if no alternatives are offered in 125 U.S. medical schools, the sole exception being the Uniformed Services University of the Health Sciences. Similar trends are developing in veterinary schools in which terminal surgery laboratories have been dropped from the curriculum.

The educational effectiveness of nonanimal alternatives

There are many studies comparing the educational effectiveness of alternatives, such as computer software and models, and they show that alternatives often are at least as good, if not better, for achieving intended educational goals. Jonathan Balcombe summarized some of these and found that for undergraduates, veterinary students, and medical students, equal knowledge or equivalent surgical skills were acquired using alternatives. The educational effectiveness of using nonanimal models was not less. For example, in a study of 2,913 first-year biology undergraduates, the examination results of 308 students who studied model rats were the same as those of 2,605 students who dissected rats. When the surgical skills of 36 third-year veterinary students who trained on soft-tissue organ models were compared to the surgical skills of students who trained on dogs and cats, the performance of each group was the same. Virtual surgery has been shown to be an effective alternative. In a study of 110 medical students, students rated computer demonstrations higher for learning about cardiovascular physiology than demonstrations using dogs. Richard Samsel and his colleagues at the University of Chicago found that first-year medical students rated both computer and animal demonstrations highly in teaching cardiovascular physiology but that the computer-based sessions received higher ratings. Many students are seeking out nonanimal alternatives.

Humans are a part of nature, not apart from nature

Roger Fouts, a psychologist who has studied chimpanzee-human communication with individuals, including Washoe, recently wrote the wonderful book *Next of Kin*. In his book. Fouts argues that by being concerned about animals and acting on these concerns a person is more of a healer than an activist.

Although not all humans believe that animals are here for us to use and abuse selfishly, or that the environment is, human arrogance prevails in many circles, and numerous animals suffer great losses in respect and immeasurable harm because of human-centered attitudes and domination of living (animate) and nonliving (inanimate) environments. In order to save the precious and fragile resources on this planet and in the universe at large, humans will have to place their own anthropocentric, selfish interests aside and work with, and not against, one another. We will also have to learn to work in harmony with, and appreciate and respect the value of, other living and nonliving cohabitants of this planet and of the universe.

There are *always* alternatives to cruelty. Humans are a part of nature, not apart from nature. Humans cannot continue to be at war with the rest of the world. The fragility of the natural order—the delicate balance of life—requires that we all work harmoniously so as not to destroy nature's wholeness, goodness, and generosity. In the absence of animals, we would live in a severely impoverished, unstimulating universe. How sad this would be. Expanding our circle of respect and understanding can help bring us all together. The community "out there" needs to become the community "in here"—in our hearts. Feelings need to be incorporated into action.

Thomas Dunlap, in his book *Saving America's Wildlife*, notes that the role of science in the development and changing of ideas is highly questionable. While many people respect scientists and bestow on them special abilities to fix things when they break, scientists and science alone will not be able to deal effectively with the many difficult and puzzling problems that arise in discussions of the nature of animal-human interactions. Personal and cultural values influence the choices we all make, and common sense also plays a large role in our decision-making. We need to figure out how facts and values are to be integrated into the choices we make about how we interact with animals and with other nature. This will not be an easy task. Open discussion about the difficult issues at hand will help us along as we try to create a world in which all life is valued, respected, and treated humanely.

Each of us holds a key to the future. Collectively, we are a powerful force. It is important to be proactive and to prevent animal abuse before it starts. Public pressure greatly influenced the use of calves for veal. Helen Steel and Dave Morris, neither of whom had any experience in legal matters, sued McDonald's in what became the longest trial in British history, the McLibel case. They successfully showed that McDonald's exploited children with its advertising, falsely advertised its food as nutritious, risked the health of its long-term regular customers, and was "culpably responsible" for cruelty to animals reared for their products. Sears, Roebuck, and Company stopped sponsoring the Ringling Brothers circus because of public pressure.

In a landmark speech delivered to the U.S. Senate in July 2001, Senator Robert Byrd (Democrat, West Virginia) railed against the rampant and barbaric institutionalized cruelty that it is inflicted on animals. Byrd condemned the widespread violations of the Humane Slaughter Act and called for the U.S. Department of Agriculture to end slaughterhouse cruelty. As chairman of the Senate Appropriations Committee, Byrd asked for an extra three million dollars for enforcement of the Animal Welfare Act and the Humane Slaughter Act. He noted that animals suffer pain and made a plea for "respect for all life . . . and for humane treatment of all creatures." Byrd also spoke about the importance of companion animals—our "unselfish friends"—for our own well-being.

Times are changing not only because more and more people, including those who have considerable influence, care about animals but also because they are actively doing something to replace cruelty with compassion. There can be no doubt that with few if any exceptions, the least invasive research will produce the most reliable and useful results. To bring about change, conviction and courage are needed. *Never say never.*

eight

◫

HUMAN INTRUSIONS INTO ANIMALS' LIVES

I now turn to the ways in which researchers studying animals can influence the behavior of the animals in which they are interested. I consider some recent examples of how various methods of study influence the animals being studied—their nesting and reproductive patterns, dominance relationships, choice of mates, use of space, vulnerability to predators, feeding habits, and caregiving behaviors. Thus general conclusions and models that are generated from these studies can be misleading because of human intrusions that appear to be neutral. I am not in any way being critical of the unintentional effects of research. In many cases it is only after the fact that we know what we did. But when we do know how we influence the very animals we want to study, we are obliged to factor this into all future research. Consequently the results will be more reliable and more ethically defensible.

While I have thought for a long time about some of the questions that are raised in this chapter, not until I began working with Dale Jamieson, then a philosopher at the University of Colorado, who relentlessly asked difficult questions about how field (and laboratory) research was conducted, did I do some soul-searching about my research methods and those of my colleagues. Dale wanted to know why some studies were done the way they were, and if he was not satisfied with my answers he wanted to know why they could not be done in other ways so as to make the data more reliable and the methods less intrusive. As a result of a long collaboration with Dale, I truly believe that my science is "harder" and "better" in that I feel confident that my results are at least as reliable as they had previously been, if not more so, and that I have not compromised the lives of the animals I was fortunate to be able to study.

In addition to making science more reliable, there is also a practical side to paying attention to how we go about studying animals. Those who study behavior and behavioral ecology, especially in the field, are well positioned to make important contributions to maintaining the well-being of the animals they study and others. Unfortunately, they often play only a minor role in informing legislation on matters of animal protection; however, they can help to provide guidelines concerning dietary requirements, space needs, the type of captive habitat that would be the most conducive to maintaining the natural activity budgets of the animals being held captive, data on social needs in terms of group size and age and sex composition, and information about the nature of the bonds that are formed between animals and human researchers. While often it is fairly easy to predict some of the effects of various methods, at other times it is not. The results from clever field studies are often counterintuitive, but nonetheless these experiments are important to conduct. For example, Guyonne Janss and his colleagues in Seville, Spain, wanted to know if the use of raptor models would reduce collisions with power lines by other birds. They used a statue of a golden eagle and found that this model provoked attacks, rather than avoidance, so was not a good way to deal with the issues at hand. We would not have known this if this study had not been done.

It is such a gross understatement to claim that humans are all over the place that I hesitate to mention it. But it must be brought to the surface, for people inevitably have significant impacts on other human, animal, and plant communities, water, the atmosphere, and inanimate landscapes. We intrude wherever and whenever we desire, and the effects of our intrusions on the lives of others have been devastating. Even where there are no human settlements, the indirect effects of human intrusions are felt by other humans, animals, and entire ecosystems. Those who study animals are not immune from this "accusation," but this does not mean that they are intentionally being intrusive. Indeed, there are many reasons why researchers have to try as hard as they can not to influence the animals they are studying if they want to learn about their "natural" behavior.

Why study animals?

Perhaps the most fundamental question regarding field (or any other) research is "Why do it at all?" Even the least invasive research can be disruptive and costs time and money. Many people study animals for deeply personal reasons—they like being outdoors, they like animals, they do not know what else they would do with their lives—but this hardly amounts to a justification. Several other reasons for doing this research are also frequently given: that animal research benefits humans, that it benefits animals, and that it benefits the environment.

Animal research that benefits humans falls into two categories. One category includes research that contributes to human health; the other category includes research that provides economic benefits. Little field research on animals can be defended on the grounds that it contributes to human health. Animal models for human diseases and disorders are better constructed under laboratory conditions, and even then many of them are extremely controversial on both scientific and moral grounds. Animal research that contributes economic benefits often concerns the control of predators. Much of this research employs morally questionable methods such as trapping, poisoning, or shooting "problem" animals.

The idea that behavioral research benefits animals and the environment is very appealing. The thought is that only by studying the animals will we know how to preserve them, and only by preserving them can we protect the natural environment. As noble as these sentiments are, they are rife with dangers, for this attitude can lead very quickly to transforming science into wildlife "management," and wildlife management poses many moral and practical challenges.

Researcher effects

> Recording and reporting such measures should be a routine part of any study using intrusive techniques, as the onus is on fieldworkers to show that their methods have no impact, or at least an acceptable impact, on their study animals.
>
> —Karen Laurenson and Tim Caro,
> "Monitoring . . . Free-living Cheetahs"

The representative studies I have chosen are meant to show how widespread researcher influences are and to highlight the diversity of species that are affected. When behavior and activity patterns are used as the litmus test for what we call normal species-typical behavior and normal patterns of variation, we need to be sure that the behavior patterns being used truly are an indication of what individuals of a given species typically do in the situation in which they are being studied. If the data are unreliable, then it is likely that the conclusions that are reached and the models that are generated are also unreliable and can mislead future research.

While some scientists are uncomfortable delving into the effects of re-searchers on the animals they study, it truly is in the best traditions of sci-ence to ask questions about methodology and ethics. It is not antiscience to question what we do when we study other animals.

Just as there are many ethical concerns in studies of captive animals, there are ethical issues associated with the study of wild animals. "Just be-ing there" can have enormous impacts. Nonetheless, field studies contribute information on the complexity and richness of animal lives that is very use-ful to those interested in matters centering on animal protection. So we need to do the best research we can in as unobtrusive a manner as possible. Needless to say, much more work and discussion is needed to flesh out just what "an acceptable impact" consists of. It is likely that what is acceptable for one species or for certain individuals will not be acceptable for others. It is possible that certain studies will have to be postponed until less intrusive and more reliable methods are developed.

It is not always the case that human activity influences animals, and these data are as important to heed as are data that show that humans do have an influence. For example, human activity was not observed to influence the growth rates of short-tailed shearwater chicks living on Great Dog Island in Bass Strait, Australia. John Byers discovered that handling young prong-horn did not increase their mortality.

Often results are site- and time-specific. Thus, while many researchers have discovered that habitat fragmentation that results in more edges can lead to higher predation along edges (called the "edge effect"), where prey are more vulnerable, this was not the case in a study conducted in Hungary and Sweden by Andras Báldi and Péter Batáry. Generalizing from one site to another and from one season to another can be risky. Comparative stud-ies conducted in different locales, at different times of day, and at various stages of the life cycles of the same and different species are needed. Ani-mals in different stages of their reproductive cycles respond differently to intrusions, and predation pressure also varies depending on whether a species is more likely to be preyed on by predators who use visual, auditory, or olfactory cues or some combination of these stimuli.

Knowing who's who

Researchers who study animal behavior often want to identify individuals, assign gender and age, follow individuals as they move about, or record var-ious physiological measurements including heart rate and body tempera-ture. Animals living under field conditions are generally more difficult to study than individuals living in more confined conditions, and various methods such as trapping, marking, and fitting telemetric devices are often used to make individuals more accessible.

Marking animals with tags or bands so that individuals can be reliably identified over time influences their behavior. Placing a tag on the wing of ruddy ducks leads to decreased rates of courtship and more time spent sleeping and preening. Data on mating patterns, activity rhythms, and maintenance behaviors from tagged ruddy ducks would be misleading. In frigate birds and boobies, banding can lead to foot and leg deformities that influence behavior.

The large effect that a colored band can have is shown clearly in small birds called zebra finches. In zebra finches, mate choice is influenced by the color of the leg band used to mark individuals, and there may be all sorts of other influences that have not been documented. Females with black rings and males with red rings have higher reproductive success than birds with other colors. Blue and green rings are especially unattractive on both females and males.

In order to follow identified individuals, humans often capture animals, handle them, and place a radio-collar around their necks or implant a telemetric device that transmits various types of information, including heart rate and body temperature.

Patterns of looking for food (foraging) can be affected by human intrusions. The foraging behavior of Little penguins (average mass of 1,100 grams) is influenced by their carrying a small telemetric device (about 60 grams) that measures the speed and depth of their dives. The small attachments result in decreased foraging efficiency. Changes in behavior such as this are called the "instrument effect" because of the influence that the "instrument," in this case a telemetric device, has on behavior, in this case diving. However, when female spotted hyenas wear radio-collars weighing less than 2 percent of their body weight, there seems to be little effect on their behavior. Similar results have been found for small rodents, for whom small radio-collars do not increase the risk of predation by birds.

When radio-collars are used, each gives off a unique signal of a specific frequency that is picked up on a receiver. Radio-collars lower breeding propensity in emperor geese. In one study of the effect of radio-collars on San Joaquin kit foxes, the researchers discovered that collared foxes showed a loss in body mass and also suffered from reduced survival. In green lizards, capture and handling can influence natural movement patterns. The information generated by studies in which animals are forced to wear radio-collars can generate misleading information on individual movement, survival patterns, and longevity. It is essential for researchers to know this when they attempt to answer questions about ranging patterns or how long individuals live.

Capturing individuals in order to place a radio-collar on them to mark them, or to weigh them, can also have an effect on other behavior patterns. Capturing and recapturing large gray mongooses influences their use of space. Thus it is important to ask if the use of space by individuals who are trapped and handled really is the use of space by individuals avoiding traps or human observers or if it indicative of how individuals use space when they are left alone. If cages in zoos are being designed to take into account animals' movement and activity patterns, then the data that are used to make decisions about designing enclosures need to be based on information that reliably indicates what the animals typically do and need in the wild. In one study of western lowland gorillas kept in zoos, it was discovered that habitat use varied between captive and wild individuals.

The weight of devices used to mark animals can also have a significant effect on their social behavior. For example, the weight of radio-collars can influence dominance relationships in adult female meadow voles. When voles wear a collar whose weight is greater than 10 percent of their live body mass, there is a significant loss of dominance. Here, erroneous data concerning dominance relationships would be generated in the absence of this knowledge.

Karen Laurenson and Tim Caro analyzed the long-term effects of wearing a radio-collar, aerial radio-tracking, and lair examination in wild cheetahs on the central plains of the Serengeti National Park, Tanzania. They found that "females wearing collars weighing less than 2% of their body weight reproduced regularly, had equivalent food intake and hunting success and were in similar body condition to uncollared females." Aerial radio-tracking "did not appear to disturb habituated females or cause them to abandon their cubs. Entering lairs on foot to count and weigh cubs while the mother was absent did not appear to increase the likelihood of cub predation by other carnivores or abandonment by the mother." Laurenson and Caro concluded that "the behaviour and reproduction of even sensitive mammals need not be affected by field techniques." However, they caution that some of their measures might have been too crude and emphasize that a lot of research is still needed to assess how various field techniques influence wild animals. Laurenson and Caro also note that there might be individual differences in response to stress (for example) that demand close attention: "Important subtleties or rare instances of adverse reactions may, therefore, have been missed."

In some instances, very simple changes in field methods and equipment can play a large role in making researchers' presence less intrusive. In a long-term study of coyotes that I conducted with Michael Wells in the

Grand Teton National Park outside Jackson, Wyoming, we were very sensitive to how our presence influenced the behavior and well-being of the animals we were observing. We found that shiny cameras and spotting scopes made the animals uneasy, so we painted our camera bodies and spotting scopes dull black so that they would not reflect much light. Furthermore, when visiting dens we always wore the same clothes so that roughly the same odor was present and so that we presented a familiar visual image to the coyotes. The coyotes eventually adapted to our daily presence and showed none of the skittishness that characterized their behavior very early in our study.

Not only do research methods influence a wide variety of behavior patterns, but they can also influence an individual's susceptibility to infection. For example, ear-tagging white-footed mice can lead to higher infestations by larval ticks because the tags impede grooming by these rodents. Thus for researchers interested in grooming and maintenance behavior, the presence of ear tags could influence results.

The effects of "just being there": Case-by-case assessments

"Just being there" and not handling animals can also influence their behavior. For example, mere human presence can influence the behavior of magpies, large, boisterous black-and-white birds with whom many people have a love-hate relationship. Magpies who are not habituated to human presence spend so much time avoiding humans that this takes time away from such essential activities as feeding. If a researcher were interested in gathering data on feeding patterns by magpies, she would have to be sure that her presence did not change feeding patterns that are typical of the species. In another species of bird, white-fronted chats, nests that were visited daily by humans suffered higher nest predation than nests that were visited only once at the end of a typical period of incubation.

Researchers have to survey populations of animals to gather information on population size, the number of males and females, and where individuals spend their time. But this is not without cost to the animals. Adélie penguins exposed to aircraft show profound changes in behavior including deviation from a direct course back to a nest and increased nest abandonment. The overall effects due to exposure to aircraft that prevents foraging penguins from returning to their nests include a decrease of 15 percent in the number of birds in a colony and an active nest mortality of 8 percent. There were also large increases in penguins' heart rates. In this case, models concerning reproductive success and parental investment would be misleading because the methods used influenced the animals being studied.

Trumpeter swans do not show such adverse effects to aircraft. However, the noise and visible presence of stopped vehicles produced changes in incubation behavior by trumpeter females that can result in decreased productivity due to increases in the mortality of eggs and hatchlings. Data on the reproductive behavior of these birds would be misleading.

Live trapping: Psychological and physical effects

Live trapping is an activity that can be extremely inhumane, and the experience of being caught in a live trap can be incredibly painful for an animal. In order to learn about the physiological (endocrinological, hematological) and behavioral responses of captive and free-ranging red foxes to padded and unpadded foothold traps, a three-year study was conducted. Trapped foxes were "euthanized" by shooting them, and nontrapped free-ranging foxes who were used as controls also were shot to death. The researchers found that foxes caught in unpadded traps had more physical injuries to the trapped limbs than foxes caught in padded traps. There were also biochemical differences between trapped and control foxes. Trapped foxes had higher levels of adrenocorticotropin, ß-endorphin, and cortisol and lower levels of thyroxine and insulin. Trapped foxes also had higher incidences of adrenal and kidney congestion and hemorrhaging in their adrenal glands, lungs, and hearts. There was no difference in the mean time spent resisting traps during an eight-hour period between foxes caught in padded (mean of 85.4 minutes) and unpadded traps (mean of 63.8 minutes). Note that animals were allowed to resist traps for cumulative periods of over one hour. This is ethically indefensible.

As a result of this study, the researchers concluded that red foxes caught in foothold traps developed classical stress response but that none of the effects of being trapped were life-threatening. Most of the changes in trapped animals were due to resisting traps. The results of their study led the researchers to recommend, all too objectively, the use of padded traps in future work. There is *no* mention at all about the ethics of either the research that was conducted or of trapping itself, an activity that should be carefully scrutinized. That padded traps do, indeed, produce fewer serious injuries had previously been shown by many other researchers, and one wonders why this research was even necessary. In one study of free-ranging black-backed jackals in Botswana, 86 percent of trapped jackals showed lameness of the leg that had been trapped.

It cannot be stressed too strongly that even if there is little or no physical damage caused by trapping, and this rarely occurs, the individual's state of mind, his or her psychological well-being, must be given serious consideration. Even if some procedures do not produce any obvious physiological or behavioral changes, animals can be feeling intense pain and deep suffering.

Methods of trapping can also lead to spurious results. In many birds, trapping methods can bias age ratios and sex ratios of avian populations or groups. Mist nets capture a higher proportion of juveniles, whereas traps capture more adults. In addition, dominant males tend to monopolize traps that are baited with food, leading to erroneous data on sex ratios. These are extremely important results because age and sex ratios are essential data for many different researchers interested in behavior, behavioral ecology, and population biology.

What does it all mean? Beware false inferences

The examples above stem from research on mammals and birds, but there are indications that human disturbance can influence the behavior and movement patterns of many other species, including invertebrates. Researchers need to determine when the effects on behavior can actually preclude their collecting reliable data for the questions in which they are interested. If misinformation is used to design future studies or captive housing and maintenance procedures in zoos or research facilities, or is used to evaluate individual well-being, then these efforts will be flawed. This is a serious matter.

Wild versus captive animals

The consequences of being subjected to various experimental protocols may be different and greater for wild animals than the ill effects experienced by captive animals, whose lives are already changed by the conditions under which they live. This is so for different types of experiments that do not involve trapping, handling, or marking individuals. Consider experimental procedures that include visiting the home ranges, territories, or dens of animals, manipulating food supply, changing the size and composition of groups by removing or adding individuals, playing back vocalizations, depositing scents (odors), distorting body features, using dummies, and manipulating the gene pool.

All of these procedures are used in studies of animal behavior and behavioral ecology, and all can change the behavior of individuals, including movement patterns, how space is used, the amount of time that is devoted to various activities including hunting, antipredatory behavior, and various types of social interactions including caregiving, social play, and dominance interactions. These changes can also influence the behavior of groups as a whole including group hunting or foraging patterns, caregiving behavior, and dominance relationships. They can also influence nontarget individuals. In addition, there are individual differences in responses to human intrusion. All these caveats need to be considered when a specific study is being evaluated.

Erring on the side of the animals: Better safe than sorry

While we often cannot know about various aspects of the behavior of animals before we begin our studies, it is well known that our presence can influence what animals do when we enter their worlds. What appear to be relatively small changes at the individual level can have both short-term and long-term effects. A guiding principle should be that the lives of the animals we are privileged to study should be respected, and when we are unsure about how our activities will influence them we should err on the side of the animals and not engage in these practices until we know or have a very informed idea about the consequences of our acts. This precautionary principle—better safe than sorry—will serve us and the animals well. Indeed, this approach could well mean that exotic animals that are so attractive to such institutions as zoos and wildlife parks need to be studied for a long time before they are brought into captivity. Recently it has been claimed that we should take this stance even regarding insects who are able to experience pain. Professor Eisemann and his colleagues conclude that this attitude "helps to preserve in the experimenter an appropriately respectful attitude towards living organisms whose physiology, though different, and perhaps simpler than our own, is as yet far from completely understood." For those who want to collect data on novel species that are to be compared to other (perhaps more common) animals, the reliability of the information may be called into question unless enough data are available that speak to the normal behavior and species-typical variation in these activities.

Are scientists accountable?

All scientists are responsible for how their results are put to use when they are aware of how they are used. For example, if we learn about how wolves live, we are responsible for making sure that this information is not used to harm them. This is not a purely "academic" issue, because a great deal of research on animals is funded by agencies that want to reduce their (and other species') populations or control their behavior. Information about the behavior of tigers or wolves may be useful to those who simply want to make rugs out of them. Those who study marine mammals have been struggling with these issues for decade. Purely scientific information about populations, migration routes, and behavior can be used by those who are involved in the commercial exploitation of whales and other marine mammals.

One idea worth considering is that a scientist who studies a particular animal may be morally required to be an advocate for that animal in the way that physicians are supposed to be advocates for their patients. On this view, the well-being of the animals a scientist studies should come first, perhaps even before the goal of obtaining peer-reviewed scientific results. Some scientists such as Jane Goodall and Dian Fossey have exemplified this ethic, but they have had many critics from within the scientific community.

Moving onward from the "good old days"

There is a continuing need to develop and improve general guidelines for research on free-living and captive animals. These guidelines should be aspirational as well as regulatory. We should not be satisfied that things are better than they were in the "bad old days," and we should work for a future in which even these enlightened times will be viewed as the bad old days. Progress has already been made in the development of guidelines, and the challenge is to make them more binding, effective, and specific. If possible, we should also work for consistency among countries that share common attitudes toward animals; research in some countries (e.g., the United States) has been traditionally less regulated than research in other Western countries.

In this evolving process of developing and improving guidelines for research, interdisciplinary input from field workers and philosophers is necessary; no single discipline can do the necessary work alone. Researchers who are exposed to the pertinent issues, and who think about them and engage in open and serious debate, can then carry these lessons into their research projects and import this knowledge to colleagues and students. Not knowing all of the subtleties of philosophical arguments—details over which even professional ethicists disagree—should not be a stumbling block nor an insurmountable barrier to learning. Few field workers understand the details of how binoculars or radio-transmitters work, but this does not prevent them from using them. The same should be true of the nuances of philosophical tradition and argument.

Many projects are not carried out because they lack scientific merit; a lack of ethical merit is every bit as important. Of course, ethical decisions can easily put one on a slippery slope, and it is not easy to know where to draw the line. For example, Marc Hauser considers the interesting and difficult question of whether or not vervet monkey infants "cry wolf" to elicit parental care. Field data are ambiguous, and Hauser suggests that "one fruitful approach to resolving these issues would be to manipulate the physical condition of mother and offspring." (Hauser does not tell us what these manipulations would consist of, but I have no doubt that he would not pursue ethically questionable manipulations.) Obviously, there might be some experiments that would be more ethical than others, and sorting among them would be very difficult when balancing the knowledge that would be gained with the type of manipulations that would have to be done. Just because these sorts of decisions are difficult does not mean that there are not better and worse answers.

How we refer to some of our research practices also needs reconsideration. We should use words that accurately describe our actions, not euphemisms that may be intended to mask them. When animals are killed it should be explicitly said that they are killed; words like "euthanize," "cull," "sacrifice," and "collect" often are used to deflect attention from the act of killing and to give a false sense of acceptability.

Would we do it again?

It is a privilege to study other animals and to share their worlds and lives with them in this more-than-human world. As we learn more about how we influence other animals, we will be able to adopt proactive, rather than reactive, research protocols. Part of learning entails changing our practices and asking, "Would we do what we did again, or have we learned something that can make other animals' lives better?"

It is more important than ever before for students to realize that to question science is not to be antiscience or anti-intellectual, and to question how humans interact with animals is not in itself to demand that humans never use animals. Questioning science will make for better, more responsible science, and questioning the ways in which humans use animals will make for more informed decisions about animal use. By making such decisions in an informed and responsible way, we can help to ensure that in the future we will not repeat the mistakes of the past and that we will move toward a world in which humans and other animals will be able to share peaceably the resources of a finite planet.

nine

◨

SCIENCE, NATURE,
AND HEART

Minding Animals and Redecorating Nature

In reality there is a single integral community of the Earth. . . . In this community every being has its own role to fulfill, its own dignity, its inner spontaneity. Every being has its own voice. . . . We have no rights to disturb the basic functioning of the biosystems of the planet. We cannot own the Earth or any part of the Earth in any absolute manner.

—Thomas Berry, *The Great Work*

All ethics so far evolved rest upon a single premise: that the individual is a member of a community of interdependent parts.

—Aldo Leopold, *A Sand County Almanac*

Here, there, everywhere

In the last chapter I focused on the ways in which researchers unintentionally influence the behavior of the very animals they want to study. Now I turn to a bigger picture, namely, how people influence other animals, and as a result nature as a whole—how interfering in the lives of individuals has a cascading domino effect on populations, species, and ecosystems on the fragile planet on which we live. Michael McKinney has discovered that human population size is positively correlated with threat to the numbers of birds and mammals for continental (but not island) nations and that mammals suffer more losses than birds during initial human impacts. His data set is convincing; 149 nations were analyzed for mammals, and 154 were analyzed for birds.

Humans are all over the place. Give us an inch and we take a foot and then some. There are no places on Earth, in bodies of water, or in the atmosphere that are not influenced by human activities. As such we are part of nature. We are not above other animals, nor are we lesser than other animals. We also are an integral piece of nature with great responsibilities that can no longer be pushed aside for convenience or because there always will be someone else to clean up the messes we leave. Our big brains convey enormous responsibility for taking care of all of nature.

A casual glance at available information clearly shows that something has gone seriously amiss in our interactions with animals and the planet in general. The year 2000 saw Miss Waldron's colobus monkey go extinct. The disappearance of this monkey from its homes in Ghana and Ivory Coast in Africa was the first in about three hundred years. We are losing about thirty thousand species a year of the estimated ten million to thirty million existing species of animals and plants. As I was writing this chapter I discovered a report from Zoo Atlanta's African Biodiversity Conservation Program that there are fewer than 450,000 African great apes still alive. Poaching and the pet trade are threatening the survival of Guinea's chimpanzees, including a group that has been studied for years and in which chimpanzees use stone tools to crack nuts.

Eastern lowland gorillas in the Kahuzi-Biéga National Park in the Democratic Republic of Congo have experienced a 50 percent reduction in numbers, and elephants are virtually gone, because of mining for coltan, a mineral that is used by Western high-tech companies to make cell phones. It can easily be mined with a shovel. Over five hundred million cell phones were sold in 2001. Cell phone users can make a difference in the amount of coltan that is illegally mined by refusing to buy phones in which this mineral is used. Indeed, the United Nations has encouraged people to boycott companies that use illegally mined coltan and other minerals.

And there is shahtoosh. Shahtoosh is a high-quality wool from the neck hair of endangered Tibetan antelopes known as chiru. This fiber is about seven times thinner than human hair, which makes shahtoosh products extremely fine and light. Shahtoosh trade is in violation of the Convention on International Trade in Endangered Species of Flora and Fauna (CITES), yet it flourishes. This illegal activity would not continue if people stopped buying these products.

The story continues. Recovery from the well-known 1989 oil spill in Alaska's Prince William Sound is weak. Of seventeen seabird species that were affected, only four show very weak recovery from the spill, and nine—comorants, various gulls, grebes, terns, and murres—show no evidence of recovery. Exxon, which was to blame for the spill, claims that "the environment in Prince William Sound is healthy, robust and thriving." A recent analysis of ecosystems in the Canadian Rockies by Charles Kay and his colleagues indicates that these analyses lack ecological integrity. These researchers studied historical records of the occurrence of such ungulates as bighorn sheep and elk and found different patterns of abundance now than in the time before European settlers arrived. The loss of carnivores and decline in hunting by native peoples has led to an increase in the abundance of these ungulates. In Colorado, acid rain, which reduces the amount of critical trace elements in the plants that bighorn sheep eat, may be the cause of the decline in sheep numbers. Global warming seems to be having an impact on the number of primates living in Ethiopia. A rise in temperature leads to less grass and fewer crops on which gelada baboons are able to graze, and it is feared that their numbers will fall because of this.

Activities such as mountain climbing also are intrusive. Climbers can influence activity patterns of birds such that they fly more and perch less and consequently waste energy. In a study of the effect of climbers on grizzly bears in Glacier National Park, Montana, researchers discovered that climber-disturbed bears spent about 50 percent less time foraging and about 50 percent more time avoiding climbers.

There are many examples of how human intrusions, including the presence of roads and traffic, influence the use of space in numerous animals. Roads influence behavior by causing changes in movement patterns, home ranges, territories, patterns of escape, and reproductive success. Roads also lead to increased hunting, fishing, and harassment of animals. Recall that roads built by logging companies have led to a large increase in the bushmeat trade. Chellis Glendinning presents some very unsettling statistics about the paving of the North American continent. In her book *Off the Map* she notes: "For every $1 million spent on the $27 billion system, these men [road builders] use 16,800 barrels of cement; 694 tons of coal; 485 tons of pipe; 76,000 tons of sand, gravel, and crushed stone; 24,000 pounds of explosives, 121,000 gallons of petroleum products; 99,000 board feet of lumber, 600 tons of steel." Energy consumption is at an all-time high but is not equally distributed among nations. For example, in 1998 North America, with about 7 percent of the world's population, accounted for nearly 30 percent of total world energy consumption. Eighty-five percent of the world's wealth is held by about 20 percent of the world's people, and 1.5 billion people live in poverty worldwide.

A bunny, two bikes, and many cars

A few years ago, when I was out for a bike ride with my friend Brad, much of what I had been feeling came home to me once again. Just as we were leaving town, a small bunny ran in front of us and dodged oncoming auto traffic. The bunny was so small that she could not jump on the curb, so she immediately ran back out into traffic. We stopped where we were, laid our bikes down in the middle of the two-lane road to halt traffic, and chased and caught the bunny. I placed her gently in the back pocket of my cycling jersey and could hear low purring sounds. All the while numerous motorists were blasting their horns at us and yelling that we should get out of the way, leave the bunny alone, for they were in a hurry—most likely to get to a job they did not like in the first place. Would ten seconds really make a difference? Just as Jethro had saved a bunny years ago, we took the bunny to a field and let her go. We do not have any idea if she survived, but surely she would not have survived her harrowing tryst in traffic. Yes, many people can be selfish and push aside any being or any thing that gets in their way. But there is a price to pay for this sort of distancing, namely, an erosion of spirit and soul.

Twelve dead and silent wolves

Three wolves in Denali National Park, Alaska, died after being darted so they could be fitted with radio-collars so that researchers could follow them. While I would never have heard their melodious howls, I know that others would have immensely enjoyed listening to them revel. If no one ever heard them, so be it, for we could delight in knowing they were "out there." Now they will be silent. Why can't we just let them be?

While I am saying a prayer for the wolves, I happen on an article informing me that in Norway a controversial wolf hunt ended with nine out of ten wolves being killed by state-appointed hunters in helicopters and snowmobiles. The hunt cost taxpayers about $36,000 per wolf. Hunting from helicopters is normally banned as unethical in Norway. "Oh, great," I say to Jethro, who is asleep on the floor of my office, "another display of human arrogance and selfishness." Jethro casually raises his head and seems to ask, "Why are you surprised?"

Luna the dolphin

The story of Luna, a bottlenose dolphin, exemplifies the arrogance with which many humans attempt to dominate other animals. Luna and seven other dolphins were captured in Magdalena Bay on the Pacific coast of Baja California, Mexico. The dolphins were transported in a truck to a small pen at the La Paz Dolphin Center to become tourist attractions.

Luna arrived in a wooden crate, exhausted, frightened, and soaked in her own blood. She and her friends were kept in a narrow pen in water as shallow as forty-five centimeters. Luna died, and her death resulted in a moratorium in Mexico on the capture of dolphins in national waters. Local activists were essential in helping out other dolphins. Project Luna was organized by Mexican environmentalist Yolanda Alaniz to close the La Paz center, and Mexican environmental minister Victor Lichtinger introduced emergency legislation to make sure that no other marine mammals shared the fate of Luna and her companions.

Drilling in Alaska: Nature as friend or nature as resource?

Humans use resources as if they are all renewable. They are not. We will surely deplete some essential resources in our lifetime or in the lifetime of our children. We have a number of choices. We can be more frugal, or we can continue to use energy recklessly and not worry about future generations. If we choose to continue being as wasteful as we have been, we will either need more synthetic fuels or be committed to drilling for fossil fuels and natural sources of energy. The latter situation raises many questions. Should we continue to harm the planet—animal and plant communities—to seek out fuel? What right do we have to do this? *Should we even ponder the possibility of continuing to deface the Earth, or should we accept that we must stop these practices, use less energy, and not continue to dominate animals and landscapes?* Along with many others, I prefer to take the latter route and to raise consciousness by having people change their perceptions so that nature is seen as a friend—a family member—rather than as an expendable resource. The vice president of the United States, Dick Cheney, disagrees. He has claimed: "The aim here is efficiency, not austerity. Conservation may be a sign of personal virtue, but it is not a sufficient basis for a sound, comprehensive energy policy."

The Arctic National Wildlife Refuge (ANWR) is a roadless area seventy-seven thousand miles square in northeastern Alaska. The ANWR is home to many species such as polar bears, grizzly bears, Dall sheep, wolves, wolverines, lynx, porcupine caribou, golden eagles, and thirty species of shorebirds. As I write, there is much debate about whether or not to explore six thousand square kilometers of the reserve's coastal plain for petroleum-based products to maintain consumptive lifestyles. Joel Berger, one of the world's leading conservation biologists, analyzed the situation on the ANWR to determine if we should even consider exploiting this pristine area, one of the world's most significant remaining intact ecosystems. He and many other experts concede that we do not know what effects exploration will have, but there is no doubt that there will be many negative influences on the lives of the animals who live there. Irreparable damage will be done to one of the world's most incredible natural treasures, and this is a significant loss to all humans. Berger and his colleagues also note that if we continue to ruin intact ecosystems, future researchers will be unable to make comparisons between natural, intact ecosystems and those in which humans have trespassed so that we can assess the impact of our intrusions. Essentially, "real" nature will be gone forever. This makes me ill.

Little is known about the intact ANWR ecosystem, and experts caution against tampering with any ecosystem until we have basic information about interrelationships among the various animals who live there and how animal and plant communities are related to one another. Retaining the biological integrity of the ANWR and other ecosystems is the main goal of conservation biologists, even if it means that some of us have to change our lifestyles and use less energy. Perhaps a useful guideline is that we should give back to the Earth what we take from the Earth. Perhaps we should give back *more* than we take to ensure a future for our children and theirs.

Redecorating nature with love: The prospect of silent seasons

> One of the great dreams of man must be to find some place between the extremes of nature and civilization where it is possible to live without regret.
> —Barry Lopez, as quoted by Bill McKibben, "An Explosion of Green"

> We're not destroying the world because we're clumsy. We're destroying the world because we are, in a very literal and deliberate way, at war with it.
> —Daniel Quinn, *Ishmael*

When humans interact with nature we frequently wind up redecorating it, selfishly. Intentionally or not, it is as if humans have a powerful inborn urge to reshape nature, to expand their horizons. It is as if we just cannot stop ourselves, and little else does, even the blatant results of our trying to dominate—manage, control—our surroundings. We move animals around as we move furniture, and we redecorate landscapes with little concern for maintaining biological integrity. Even during strolls in pristine forests, swims in oceans, or forays in the sky, many humans are detached and alienated from the majesty of their surroundings. It seems as if they do not love nature deeply and that they would not miss nature and wilderness that are disappearing at unprecedented and alarming rates.

Is there a solution for the situation we have created? In my view, holistic and heart-driven compassionate science needs to replace reductionist and impersonal science, ethological, environmental, and otherwise. Creative proactive solutions drenched in deep caring, respect, and love for the universe need to be developed to deal with the broad range of challenging problems with which we are confronted. Love also is an essential ingredient in the recipe for reconciliation. Its power must not be underestimated as we forge ahead to reconnect with nature. Passionate impatience and quick fixes need to be tempered with compassion and love. Often we sit around and ponder the crises for which we are responsible as wilderness and wildness slip away right in front of our senses.

Silent seasons are looming. We should indeed fear that silent springs may be followed by silent summers, winters, and falls. California condors, gray wolves, Canadian lynx, and numerous but less appealing and less known animals have been brought to the brink of extinction in various locales by human activities and greed. Primary forests have all but disappeared. Shopping malls and parking lots take precedence over the lives of such threatened species as black-tailed prairie dogs, whose close-knit families are decimated by bulldozers and by drowning them in their underground homes.

In his book *The Great Work*, Thomas Berry stresses that we should strive for a benign presence in nature. We can enjoy nature but not dominate or ruin it. One way to have a benign presence is to ensure that there are places that are thoroughly protected from all human intrusions. Paul Gruchow notes that there is a necessity for empty places. I could not agree more.

It is all too easy to throw up our hands and give up hope. We have truly made some egregious errors. But it is important to remain hopeful if there is any possible salvation, if we are ever to reconnect to nature as our ancestors connected during their long evolutionary history. Indeed, life as we now know it is only a blip in time. Even more disturbing than a doomsday view that the world won't even exist in a hundred years if we fail to accept our unique responsibilities is imagining a world in which humans and other life coexist in the absence of any intimacy and interconnectedness. Surely we do not want to be remembered as the generation that killed nature. Nature can be our friend, and reconnecting with nature can help overcome alienation and loneliness. I know I would miss other animals and natural environs more than I would miss a finger or two.

Science and social responsibility

I am a hopeful and optimistic scientist although I find strict scientistic attitudes to be too limiting and bereft of compassion. One solution that will likely make a big difference in helping us solve ecological crises would be to make science more socially responsible and more compassionate. While practicing scientists are the key to deep and heartfelt change, nonscientists also can have an impact by forcing science and scientists to "go public" with their findings. Arrogance and know-it-all attitudes need to be replaced with humility and honesty. We do not really have all the answers for the difficult questions that continually bombard us concerning the numerous ecological crises that are present and looming. If we did, we would not find ourselves in the situation we are in now, nor would problems about energy consumption, pollution, or imperiled species recur over and over again.

Bumper stickers often contain words of wisdom. One in particular has been percolating in my brain for many years: "Back off, man, I am a scientist." I first saw it when I was in graduate school (and later heard my college friend Harold Ramis utter the same words as the mad scientist in the movie *Ghostbusters*), and it really bothered me because I thought it sent a false message concerning the arrogance of scientists. If nonscientists truly knew what we do, I thought, they would not think we were so removed from the mainstream; they would see a more humane and less objective side to science. Now I am not so sure. Much of science is self-serving, and many scientists behave as if they do not want nonscientists to catch wind of what they really do.

I am a scientist and very proud to be one. Because I have a dream of reconnecting humans with the lives, souls, spirits, and hearts of other animals, and also with inanimate landscapes, some of my colleagues think that I am a bit bizarre and that my science is too "soft." Because I am a sentimentalist—I love animals and nature—some think that my science is flawed, too subjective, with little or no hope for redemption.

I truly love what I do. "Soft" or "hard," I "do science." But I do not take a reductionist and impersonal scientistic view of the world in and around me. I am in awe of how much nature has to offer when we take leave of our heads and open our hearts to her boundless and breathtaking splendor, her innumerable messages, her beneficence, her generous invitation to join her. But having fun, being sentimental, and doing science are viewed by some as being mutually exclusive.

Science . . .

It is often valuable to step back and take a look at whatever it is we do. As a scientist I often watch myself "doing science." Asking questions about science can be useful for learning about science and scientists. Science supposedly tells us why things are the way they are. However, science is not value-free. Numerous biases are embedded in scientific training and thinking. Scientists, as human beings, have individual agendas, personal, social, economical, and political. Science needs to be more open to individuals' worldviews. There are so many diverse problems, it is unlikely there is only one sound scientific method. A little pluralism can go a long way. Surely we cannot do much worse than previous efforts in our attempts to reconnect with nature and to confront environmental crises.

What do scientists typically do? There is a basic structure to doing science that most scientists follow, no matter how different are our interests. We ask questions, design research projects to answer these questions as unambiguously as possible, analyze data, see how well our results fit our predictions, generalize to other situations, write up papers, deliver presentations, make errors, and go back to the drawing board to design future work. Basically, science proceeds by a combination of supporting predictions, making errors, discovering new connections and patterns among variables, and then designing future projects. Scientists, like other humans, are fallible. Indeed, our fallibility keeps us in business.

For many decades, science and scientists have been held in high esteem and placed on a pedestal by nonscientists and scientists themselves. Numerous scientists had an arrogant attitude about their self-worth, an attitude that did not serve science well. Most scientists work in a safe, insulated microcosm. Scientists were trusted; their authority was unquestioned, and those who questioned it were considered to be members of fringe groups, perhaps even Luddites, who were antiscience or anti-intellectual. Scientists were generally autonomous, and a monologue generally went from science to the rest of society with little exchange or interaction. After all, scientists busily discover cures for diseases, the structure of the human genome, how to make weapons for global destruction, ways to get to the moon and elsewhere, how to generate and process information faster, how to engineer better food, how animals behave, how nature works, how to make our lives longer and presumably better. Indeed, science has chalked up innumerable successes. But surely it can do better. And we should demand that it does.

Nowadays more people, including some scientists, question science. Increasingly science is seen not as a self-justifying activity but as another institution whose claims on the public treasury must be defended. Nonscientists are generally more aware and more inquisitive, and society is more complex. They believe that there needs to be a new social contract between science and society that is characterized by two-way dialogue. Science continually has to be legitimized.

Fragmenting nature: Replacing reductionism with integrative holism

While science has allowed us to discover much useful and important information about nature, one reason traditional science often falls short is that it fragments the world. It forces a separation between the seer and the seen, how the world is felt and sensed. Reductionistic science that splits the world into little pieces sorts and filters reality, dissects, disembodies, and splits wholes into parts; it makes holes in wholes. It produces linear, mechanistic views of the universe and objectifies and devalues animals and nature. It reduces the multidimensionality of our interactions with other animals and nature into dimensionless and static flatlands rather than stimulating the development, understanding, and appreciation of mottled and contoured landscapes. After much is learned about how various components of whole systems work, and the time seems right, scientists then try to reconstruct the wholes that they have hacked apart. But unfortunately we are not very good at reassembling the wholes; we cannot put Humpty Dumpty back together once he has been dissected. Despite good intentions, we often discover that the whole is greater than the sum of its parts, and we are unable to understand how whole systems emerge from complex interdependent interactions among their constituents. The system that emerges from reconstructing the whole is a rather mysterious one.

Webs of nature: Coyotes, cats, and scrub birds

There are complex webs of nature that, when disrupted, can lead to "silence," the loss of birdsong. Kevin Crooks and Michael Soulé have a very informative story to tell about complex interrelationships among coyotes, other predators (known as mesopredators) such as domestic cats, opossum, and raccoons, and scrub birds including California quail, wren tits, spotted towhees, Bewick's wrens, California thrashers, greater roadrunners, cactus wrens, and California gnatcatchers living near San Diego, California. Their research is an example of the importance of long-term projects that investigate complex webs of nature that are not obvious at first glance. Crooks and Soulé found that scrub bird diversity, the number of different species present, was higher in areas where coyotes were present. Domestic cats, opossum, and raccoons avoid coyotes by shunning areas where coyotes are most active, and birds benefit. The disappearance of a dominant carnivore, the coyote, resulted in elevated numbers and activity of mesopredators who exert strong predation pressure on native prey species. Coyotes kill domestic cats where they cohabit; cat remains were found in 21 percent of 219 coyote scats collected in these areas. Twenty-five percent of radio-collared cats also were killed by coyotes.

Unlike wild predators, domestic cats are recreational hunters whose high numbers are maintained by nutritional subsidies from their owners; they continue to kill birds even when bird populations are low. Thirty-two percent of residents bordering the San Diego area where Crooks and Soulé conducted their study owned cats. On average each cat owner owned 1.7 cats. Seventy-seven percent of cat owners let their cats outdoors, and 84 percent of outdoor cats brought back kills to the residence. Cat owners reported that each outdoor cat who hunted returned on average twenty-four rodents, fifteen birds and seventeen lizards to the residence each year. This is a large number of victims.

Crooks and Soulé discovered that the level of bird predation appeared to be unsustainable. Existing population sizes of some birds do not exceed ten individuals in small to moderately sized areas, so even modest increases in predation pressure from mesopredators, in conjunction with other fragmentation effects—those alterations of the environment that produce separate islands of land—may quickly drive native prey species, especially rare ones, to extinction. Extinctions of scrub-breeding birds are frequent and rapid. At least seventy-five local extinctions may have occurred in these areas over the past century.

Enjoying science, having fun, and attracting students

> Many of the students who have crossed my path in the last decade or so have wanted to do much, much more. They were drawn to ecology because they were brought up in a "world of wounds," and want to help heal it. But the current structure of ecology tends to dissuade them. . . . Now we need to incorporate the idea that it is every scientist's obligation to communicate pertinent portions of her or his results to decision-makers and the general public.
>
> —Paul Ehrlich

So claims one of the most influential ecologists of our time in a very important book, A *World of Wounds: Ecologists and the Human Dilemma.* Later Ehrlich writes: "For environmental scientists the new ethic should also include, in many cases, doing more than just trying to discover how environmental systems work. It should involve putting some time into making them work better." I was also very pleased to read the following:

> In my view, no area of science can be successful (or much fun!) without a mutually supportive interaction between theory and empiricism. . . . So let's stop arguing about theory versus empiricism and worrying about the end of our science. Instead, let's cooperate more, change some of our priorities, and have fun while we're trying to save the world.

One way to make science better is to make doing science fun. Perhaps we need to play more, blend lightness into serious science, and be less cantankerous with our opponents. Indeed, if we want students to choose scientific careers, we need to show them that it is fun, that doing science is a challenging adventure in which individual creativity is rewarded. There are many examples in the history of science of the "aha phenomenon"—when scientists see how to solve something because they have brushed aside the restraints of traditional scientific linear thinking and allowed themselves to engage in multidimensional interdisciplinary musings that are challenging and fun. Many people report that highly creative solutions come to them when they are "just out there doing something else and having fun," when they are relaxed and their thought processes are not stuck in the spasms of a cerebral traffic jam.

I have had students complain to me that they want to do interdisciplinary and holistic research because they realize that reductionist science misrepresents the world. Misrepresentation has serious consequences for the quality of knowledge we gain and for how we interact in and with nature. Reductionism promotes alienation, isolation, and disconnection, and some fear that doing this sort of ecological science would not be much fun at all, for it forces a separation between the seer and the seen and creates a false dualism. Science can make nature less majestic and less magical by impeding our truly sensing, feeling, and understanding the scope of the amazing world in which we live.

To help us along in this age of alienation, holistic and more heart-driven science would be beneficial, science that is infused with spirit, compassion, humility, grace, and love. Closet holists need to emerge and offer their heretical views. Holistic heartfelt science reinforces a sense of togetherness and relationship, family and community, and awe. It fosters the development of deep and reciprocal friendships among humans, animals, and other nature. It helps us resonate with nature's radiance and lessens our tendency to think, egocentrically, that we are at the center of everything. And I bet it would be more fun for aspiring students.

The importance of traditional knowledge

> The most interesting point made about Arctic scientists is that so few of them live in the Arctic. While Inuit knowledge is formed through "being told," "doing," "hearing about it," and "being there,"—all interactive and personalized forms of knowledge transmission—Arctic science is shaped by external factors.
>
> —Laura Nader, *Naked Science*

Laura Nader's observation is a very important one to bear in mind. Science needs to be redefined to include hard data infused with stories, anecdotes, traditional knowledge, and down-home common sense. There are many ways of knowing.

Firket Berkes stresses the importance of paying deep attention to traditional ecological knowledge by providing numerous examples of how Western science cannot deal with many "local" problems encountered in foreign lands. He notes, for example, that scientists did not know that there was a population of eider ducks that lived year-round in Hudson Bay, but the Inuits did. The Inuits' knowledge was for a long time ignored in summaries of the avifauna in this area because it was not "scientific."

Berkes also warns that visiting scientists often have a "seasonally limited research period," the result of which is that they cannot possibly learn about the long-term details needed to make substantive claims about ecological problems. In the Keoladeo National Park in India, local people argued for years that grazing by water buffalo should be allowed because it was consistent with conservation objectives. Park authorities disagreed. A long-term study by the Bombay Natural History Society supported the local peoples' claim. Grazing helped counter the tendency of the wetland to turn into grassland. A ban on grazing had negatively affected the wetland and the park, which was well known for its rich bird life. Grazing by cattle was an effective solution.

Switching gears

There are many ways to do "good" science. Plurality is important. One route would reinforce creative, passionate, and bold dreaming and resist narrow thinking that claims there is only one way to do "good" science. Allowing individual idiosyncrasies, interdisciplinary collaborations, holism, and heart and love to inspire science will make it more exciting, creative, fun, and attractive to students—and likely better. The renowned scientist Frederick Seitz has lamented how disturbing it is to learn that few scientists under fifty years of age have much interest in things outside their discipline.

What could possibly be lost if scientists were open to change, or at least allowed others to pursue their own brand of science? But some still resist the notion that science is value-laden, and some do not want to impregnate science with feeling. Some scientists do not want to blend the natural sciences, values, and social sciences because this supposedly will lead to the defeat of scientific truth. I am always pleased when I recall that the Nobel Prize–winning geneticist Barbara McClintock stressed that scientists should have a feeling for the organism with which they work, and she worked with corn!

Moving animals from place to place: Redecorating ecosystems

Moving animals from one place to another, translocating them, is very often done to help endangered or imperiled species. T. Tear and his colleagues reported that 70 percent of species recovery plans involve translocating individuals and 64 percent involve reintroducing species to areas where they once lived. They also reported that for animals there is an average lag time of 11.3 years between the time a species is listed as endangered and the development of its recovery plan. Suffice it to say, we need better predictors of when species are in trouble, for during this long period of time species numbers could become irreversibly low or a species could go extinct.

It is very clear that knowledge of the behavior and behavioral ecology of animals is essential for any and all conservation efforts. Many difficult issues arise in these projects that demand careful attention before they are implemented. Redecorating nature with live and sentient beings is not as easy as redecorating one's living room. Careful proactive planning is essential. There is also a need for making informed decisions as a project progresses, for there are many unanticipated effects of moving species about. Nonetheless, in some highly visible projects there has been little proactive planning and a good deal of trying to minimize or control damage because of poor planning. Proactivity entails such tasks as evaluating critical habitat, determining public attitude, and assessing the impact of moving animals from one place to another not only on the animals themselves but also on other animals in the habitat where they are placed and on the ecosystem from which they were removed. There are cascading effects when animals are moved about, but usually there is too much emphasis on the location to which animals are moved rather than on the possible holes that are left in their former homes. For example, when such important "keystone" predators as gray wolves are moved around, there are effects on coyotes, foxes, mountain lions, bald eagles, and rodents.

One difficult question that arises in translocation studies is "Should individuals be traded off for the good of their species?" Although conservation biologists need to deal with individuals, populations, species, and ecosystems, they usually are more concerned with populations, species, and ecosystems. However, some conservation biologists are troubled when making decisions about the relative value of individuals versus species, populations, and ecosystems. It is worth quoting Jim Estes, a very experienced conservation biologist, on this issue, for he poignantly and succinctly gets to the heart of the matter in his discussion of whether or not to rehabilitate oiled wildlife, specifically California sea otters.

> The differing views between those who value the welfare of individuals and those who value the welfare of populations should be a real concern

to conservation biology because they are taking people with an ostensibly common goal in different directions. Can these views be reconciled for the common good of nature? I'm not sure, although I believe the populationists have it wrong in trying to convince the individualists to see the errors of their ways. The challenge is not so much for individualists to build a program that is compatible with conservation—to date they haven't had to—but for conservationists to somehow build a program that embraces the goals and values of individualists because the majority of our society has such a deep emotional attachment to the welfare of individual animals. . . . As much as many populationists may be offended by this argument, it is surely an issue that must be dealt with if we are to build an effective conservation program.

Clearly, if individuals are favored, then there is little hope for long-term success of certain types of reintroduction projects, for it has always been the case that some individuals will die during the experiment.

Many of the main issues concerning trade-offs among individuals, populations, species, and ecosystems are highlighted when considering reintroduction programs. Some of the above ideas about integrative, compassionate, and holistic science also find a home in conservation biology. Moving animals from one place to another, translocating them and redecorating various habitats, also raises many questions concerning humans' relationships with nature. What role should humans play in managing and controlling nature? Should we even try to restore or recreate ecosystems? Is increasing biodiversity good—is more better? These big questions require broad and interdisciplinary, rather than narrow and reductionist, views of science. In addition, sociological, economical, political, and biological aspects and agendas demand close attention, for in the end, it is humans moving around nonconsenting animals and changing ecosystems.

Translocation projects involve capturing animals in one area and transporting and releasing them elsewhere. These events are psychologically and physically stressful to the animals who are moved about. Furthermore, the ecosystems from which the animals are taken and the ecosystems into which the animals are placed undergo changes, but there have been few studies of what happens in each area after animals are removed from or inserted into a given area. When research is conducted, it invariably focuses on the fate of the introduced animals in their new homes.

Jinxed lynx: A case study

A recent attempt to reintroduce Canadian lynx into their historical range in southwestern Colorado stimulated me to write an essay titled "Jinxed

Lynx?" in which I raised many questions that center on the complex relationships between humans and other animals. Lynx previously lived in various areas of Colorado but were very rare or, according to some biologists, nonexistent when the reintroduction project began. Lynx are now listed as "threatened" under the Endangered Species Act.

In Colorado, during the winter and spring of 1999, forty-one Canadian lynx were reintroduced to areas where they once roamed. Another fifty-five individuals were released in April 2000. Four of the lynx released in early 1999 died of starvation soon after they were freed. As of November 2001, thirty-nine lynx had died, including nine who starved to death, five who were shot, and five who were hit by cars. There were forty-four known survivors, and the whereabouts of twenty-two were a mystery. There is no indication that there has been any successful breeding, the absence of which spells doom for the future of Colorado lynx.

This highly controversial project brings to light some concerns about reintroduction efforts and humans' role in trying to control and redecorate nature. For example, it is not clear that species preservation and conservation *have* to be valued, why "more is better," why biodiversity should be conserved, or if we can truly improve nature. Just because we can do something does not mean we *ought* to do it. There are numerous factors beyond the control of scientists and others who so dearly want reintroduction projects to succeed. Some biologists argue that personal attitudes, human shortsightedness, and greed would, with few exceptions, be insurmountable stumbling blocks in attempts to manage animal populations.

I do not raise these questions because I am against all reintroduction and translocation programs. Indeed, some well-planned efforts look to be on the road to yielding sustaining populations (gray wolf recovery in Yellowstone National Park seems to be progressing faster than predicted, and red wolves are doing well on the Alligator River National Wildlife Refuge in northeastern North Carolina), and they can serve as models for future efforts. I ask these questions because the issues are not as clear as some people want them to be. The Yellowstone project was well motivated and may be "successful" if it continues to proceed as it has, whereas the Colorado project was more "decorative" in that the effect of the lynx on the ecosystem was not a motivating factor. While I deeply appreciate the good intentions and efforts of all involved, sometimes good intentions are not enough. There is no room for failure, for these highly visible projects continually come under careful public scrutiny.

African wild dogs: What really happened?

African wild dogs are highly social canids whose survival is closely related to cooperative social dynamics (breeding, caregiving, and hunting) among pack members. A highly visible and disputed example of the possible effects of human interference into wild populations concerns the plight of these African wild dogs. Interference into the lives of wild dogs involved vaccinating them against rabies and canine distemper. While some scientists maintain that handling the dogs and inoculating them was indirectly responsible for their decline because the handling weakened the dogs' immune system, making them less resistant to stress, others, using the same data, conclude just the opposite. Here we have an example of scientists, all of whom care deeply about African wild dogs, not being able to discern what caused their decline. This is because the problems are so incredibly difficult. Should the researchers interfere and possibly cause animals to die, or let nature take its course? If the rabies and distemper were introduced by domestic dogs who would not have been there in the absence of people, are we more obligated to try to help the wild dogs than if the rabies and distemper were natural?

Clearly, human presence poses a serious threat to wild dogs. Rosie Woodroffe and Joshua Ginsberg, two outstanding conservation biologists, concluded that "61% of recorded adult mortality is caused directly by human activity." However, correlation and causation can be conflated in studies of the effects of human intrusion. Thus Ginsberg and his colleagues concluded that disease was responsible for the collapse in the population of wild dogs living in the Serengeti-Mara region and that handling was correlated but not causally related to mortality. Marion East and Heribert Hofer, two very experienced field workers, disagree. The case is still open.

James Kirkwood, at the Universities Federation for Animal Welfare in England, has considered such questions as whether we should intervene on behalf of free-living wild animals, and if so, to what extent and how it should be done. While Kirkwood recognizes that there are many different views, he claims that "most would probably agree that when wild animals are harmed by man's very recent (in evolutionary terms) changes to the environment (such as oil-spills, power lines, roads, and environmental contamination) there is a reasonable case, on welfare grounds, to intervene." Kirkwood calls for "an international code on intervention for wildlife welfare to provide guidance on ethics, methods and standards." This is a very good idea.

Are we faking nature?

Nature is complex, but many people want simple, quick solutions when tinkering with her. There are few, if any, quick fixes. Successful proactive planning takes time. When trying to conserve species or restore ecosystems, we must be concerned with all individuals who are involved, not only human-centered goals. Many lives are at stake, and there are many "big" questions. Should individuals be moved and perhaps suffer and die because of what we want? Should individuals be traded off for the perceived good of their species? Should populations and ecosystems that have developed and sustained themselves in the absence of predators be altered? What about other predators who might now experience increased competition for food? For example, reintroduced wolves and their offspring are killing numerous coyotes (estimated by Bob Crabtree and Jenny Sheldon to be more than 50 percent of the coyote population) in the Lamar Valley in Yellowstone National Park. In areas of high wolf activity the reduction in coyotes is as high as 80 percent to 90 percent, whereas in areas of low wolf activity the reduction is about 30 percent to 50 percent. Before wolves were reintroduced, extant coyotes did not compete with wolves. Why are wolves more valuable than coyotes? Are they? The implication is that wolves are more valuable, for no one has tried to stop wolf reintroduction because the wolves are killing coyotes. What about prey who now will be eaten when in the past, in the absence of wolves, they would not have been preyed upon?

It may turn out in some cases that it is simply impossible to regain what was lost. It may be infeasible to re-create what once existed, because times have changed. In the end, perhaps we are simply faking nature by trying to re-create her.

Moving animals around also is related to arguments about the necessity of adopting integrative and holistic views of science and nature. Removing individuals involves dismantling an ecosystem, and when individuals are removed we change the relationships among remaining components. When we introduce individuals into an area, we also change relationships among variables at this location. A key question centers on how we deal with the new properties that emerge at both locations. This is not a trivial question nor an easy one to solve. We deal with it every time we redecorate nature. Reductionist views are unlikely to help us understand what we have done when many interacting variables are involved.

Minding animals: Coming full circle

One of the best ways to reconnect with nature is to study the behavior of the awesome and mysterious animals with whom we share this splendid planet. A deep understanding and appreciation of our animal friends, who they are in their own worlds, will help them and us along, and that is why I stress how important it is to "mind" animals. I love my animal friends who so generously share their worldviews with me as I walk around my mountain home or study them as they play with one another, watch out for predators who are looking for their next meal, or lovingly protect and care for their young. I recall with much warmth the encounters with wildlife that I had with my father five decades ago, how he and I were overwhelmed by the beauty and splendor of a small and lithe red fox as he ran in front of us while we were cross-country skiing in Pennsylvania.

The diversity and mystery of animals' lives and the individual differences in their personalities are exciting to study and to ponder, and each and every one of these beings is magical and awe-inspiring. Trying to figure out what is happening in their minds keeps me on my toes and demands that I give them my full attention, that I be mindful of who they are. I cannot imagine spending my time doing anything else, and I feel fortunate and blessed to be able to make my living watching animals do what they do. In many instances we can learn about our animal kin by simply watching them as we go about our daily routines—sitting at the dining room table, walking with our dogs, or riding bikes or hiking. Watching a mother robin feed her squealing young or seeing two dogs greeting one another as pals with their tails whipping about like windmills is all I need to make a good day an even better one. Animals continually help me deal with the "big" question of who I am in the massive but interconnected universe. Reconnecting with nature can help overcome alienation and loneliness.

Solid science can be driven by our fine-tuned heartstrings. Solid science can be done even if one marches to the beat of a different drummer, for there are many ways of knowing. Saturating science with spirit, compassion, humility, grace, and love can help bring science and nature together into a unified whole. As a result, magnificent nature, the cacophony of her deep and rich sensuality, will be preserved and respected, cherished, and loved.

ten

ANIMALS AND THEOLOGY

Stepping Lightly with Grace, Humility, Respect, Compassion, and Love

We are not alone on this planet, even though our behavior at times suggests otherwise. The manic pace of our modern lives can be brought into balance by simply giving in to the silence of the desert, the pounding of a Pacific surf, the darkness and brilliance of a night sky far away from a city. . . . Wilderness is a place of humility. Humility is a place of wilderness. . . . The eyes of the future are looking back at us and they are praying for us to see beyond our own time . . . that we might act with restraint, that we might leave room for the life that is destined to come. . . . Wild mercy is at our hands.

—Terry Tempest Williams, *Red*

My experiences have been a kaleidoscope of conducting "hard science," watching and living with animals around my mountain home, stumbling over mountain lions, talking to people who know next to nothing about science, engaging adversaries, and engaging myself—literally catching myself in the act—as I try to figure out what was happening at any given moment. I have freely trespassed into many different fields, for I find them, and the numerous and deep connections among them, to be fascinating. I have tried to present solid science and to ground my speculations in science, heart, grace, feeling, and love. From now on it will not be business as usual for science. Compassionate and socially responsible science are emerging and are sorely needed as we forge ahead to reconnect with other animals and to heal the wounds we have inflicted on Earth. We need animals, and we need wildness and wilderness, to be healthy human beings.

We also need to admit past mistakes and step lightly as we move forward with boundless generosity and kindness toward all beings. A science of reconciliation and heart will help us immensely.

Pluralism and interdisciplinary and open discussions are essential. Science cannot do it alone. Nor can individuals working in isolation from others. Holism must replace fragmented science and academic territoriality. We need to allow animals to be our teachers and healers. Anthropocentrism needs to be replaced with more heartfelt biocentrism and egalitarianism. We must develop a new paradigm of compassion, respect, and love for all animals and our wondrous planet. These are tall orders. The philosopher Mary Midgley has recently argued that science and poetry can be compatible bedfellows, that subjectivity is not scandalous, that holism is the wave of the future, and that there is unity to our lives. She notes, "We can resist the academic fashions that now fragment us." Not only *can* we resist them, but we *must* for better science and for better tomorrows.

As I write I am at once smiling and feeling twinges of sadness, for I have touched on so many topics and much more work needs to be done. It is unsettling that nearly one-half of our splendid planet has been transformed so that there are "dead zones," areas where there is little or no oxygen in coastal waters. And we only have one Earth.

But I am an unwavering optimist, and I hope that I have whetted your appetite for more in-depth explorations on any or all of the topics about which I have written. We need to keep our curiosity alive and well. I find that being insatiably curious makes me feel good, and feeling good leads to more optimism, hope, and love. This cycle keeps repeating itself with no clear beginning or end.

Embodiment and emergence: Blurred boundaries

> Surely it is our animal nature that recognizes the divinity of the natural world in all its mystery and beauty, despite the distressing habits and limited perception that afflict our species. So perhaps our hope of redemption lies in the fact that we are animals, not that we are people.
> —Elizabeth Marshall Thomas, *Certain Poor Shepherds*

Interacting closely with many animals on their own terms has had a large effect on my own reflections on my place in the universe. These intimate encounters have naturally led to questions of self-identity and human uniqueness. I cannot resist the urge to try to tie together the loose strands of thinking and feeling. My quest has been daunting and stilling, and it has led me into terrain into which I never thought I would venture.

Who am I? What am I? Where am I? When am I? Asking these questions is a useful way to begin a brief discussion of my spiritual quest in which I bring together my brand of science—ethology, the study of animal behavior—and spirituality. Learning about other animal beings—how they spend their time, who they interact with, where they do what they do and how, their intellectual and cognitive abilities, and their deep emotional lives—is essential for gaining a full appreciation of human spirituality and what it is to be human.

Who am I? What am I? Where am I? When am I? We have all thought deeply about these questions, and at least for me (as far as I can parse who "me" is) the answers seem to change daily. The various iterations, at once unnerving, intriguing, amusing, and always mysterious, only make me want to know more about myself and the world in which I am presently situated. The various editions of the answers to these questions drive me deeper and deeper into my own spirituality, into my own humanness, in which tangible and intangible elements reside, resonate, radiate, and freely associate in the existential milieu. No matter what the temporary answers turn out to be, it is always a rich sense of unfettered amazement and awe that ignites my mind and fuels further travels into the much welcomed shadowy soup. If it were a smooth journey, if my straying were easy, there would likely be something amiss.

When people ask me who I am, I tell them that I am a human being first and an ethologist second. I do not define who I am by what I do (although I do love what I do, for it is a lot of fun). I have discovered that animals truly play a large role in my defining my place in the world. As I journey into spiritual dimensions and use knowledge of the lives of other animals in their own worlds as a guide, I discover much about my own nature and about human nature. I can only assume that some of what I discover about my own place in this awe-inspiring, mysterious, and wondrous world is related to the experience of other human beings. My own travels—spirited adventures for the most part—have been at various times smooth, difficult, and troubling, but they have been delightfully and incredibly rewarding and transforming, for they have opened the doors of my heart and all of my senses to the lives of other animals.

I have discovered and reconfirmed that my own nature and spirituality (and consciousness and sociality) as a transient visitor to Earth are dynamically embodied in who I am and also are defined by ever-changing emergent relationships with other beings. All beings are defined as a combination of who is "in here," in their own hearts and heads, and who is "out there," in the social matrix of the external world. I am a mysterious social phenomenon, who does not live in isolation from others, animate or inanimate. I am an embodied and emergent social phenomenon.

The social matrix in which I am defined is an integrated tapestry, a dynamic event of monumental proportions that may currently (or forever) resist being totally intelligible given the evolutionary state of my brain (and other humans' brains). My spiritual quest has taken me to the arena in which science, ethology, spirituality, and theology meet. Much of my journey owes itself to my interactions with other animals and their willingness to share their lives with me. Watching a red fox bury another red fox, observing the birth of coyote pups and the tender care provided by parents and helpers, watching dogs blissfully lost in play, and nearly stepping on a mountain lion make me realize how much of "me" is defined by my intimate interrelationships with others.

"Dead dog walking": A spiritual ritual that makes us human

Our companion animals are very special to us. While relatively few people ever get to experience wild animals in nature, many develop close relationships with companion dog and cats. It is when we have to take the life of another being, the ritual of compassionately euthanizing them or "putting them to sleep," that who we are in the grand scheme of things comes to the fore. There is no evidence that nonhuman animals choose to end the life of another who is suffering. Choosing to end another's life, being responsible for someone's last breath of air, is a deep lesson in healing, spirituality, and love. The ring of "dead dog walking" carries with it many important messages that help define our place in the world.

The importance of unwavering compassion and hope

> But we humans are newcomers. . . . We have not had much experience in voluntary interspecies (or even intraspecies) cooperation. We are very devoted to the short-term and hardly ever think about the long-term. There is no guarantee that we will be wise enough to understand our planetwide closed ecological system, or to modify our behavior in accord with that understanding. Our planet is indivisible.
>
> —Carl Sagan, *Billions and Billions*

> One of the most ubiquitous aspects of the natural world is *synergy*—combinatorial effects that are produced by joint action of two or more discrete elements, components, or individuals.
>
> —Peter Corning, *The Synergism Hypothesis*

I have often wondered what science might look like if, instead of having animals in numbered lots, they were treated respectfully as individuals . . . what science might have become had its history been different, had it not relied on distancing ourselves from nature.

—Lynda Birke, *Feminism, Animals, and Science*

Humans truly are a recent addition to Earth. George Fisher, a professor of geology at Johns Hopkins University, has developed a novel metric by which to view human history in the vastness of cosmic and geological time. Imagine that a century is represented by one hundred millimeters, a millennium by one meter, and a million years by one kilometer. Imagine also that we travel only one millimeter per year. Fisher notes that the Spanish discovery of America would take only half a meter and that returning to the time of Socrates would involve a trip of two and a half meters. If the Washington Monument in central Washington, D.C., is used to mark the present, a journey back to the Big Bang would take us across the United States to Tokyo. By moving one millimeter per year we could observe the entire story of the cosmos unfold. The Earth and solar system begin to form at the coast of California, and dinosaurs appear and disappear in northern Virginia. Tool-using hominids appear at the Potomac River, merely two kilometers west of the Washington Monument, Cro-Magnon cave artists emerge about thirty meters from the center of the monument, and all of human history, from the Sumerian city-states to the present, *fits in the base of the monument itself.*

We truly are newcomers to Earth. In his poem "Please Call Me by My True Names," Thich Nhat Hanh writes: "We are the shared emotions of all our brethren, We are truly a kindred spirit with all of life." We are all part of the same deeply interconnected and interdependent community. We are one among many. We are all woven into a seamless tapestry of unity—of friendships—with interconnecting bonds that are reciprocal and overflow with respect, compassion, and love. We share a universal pulse. I feel blessed when I open myself to the heart, spirit, and soul of other animals. When I study coyotes, I am coyote; when I study birds, I am bird. Often when I stare at a tree, I am tree. There is a strong sense of oneness.

Compassion and hope are two essential ingredients for making this a better planet for all life. My own spirituality is based on a deep drive for a seamless unity, a sense of oneness, motivated by compassion, respect, and love. Alan Sponberg presents a useful model of compassion in his "hierarchy of compassion." In his hierarchy "vertical progress is a matter of 'reaching out,' actively and consciously, to affirm an ever-widening circle of expressed interrelatedness. . . . Progress along this spiral path confers no increasing privilege over those who are below on the path. Quite the contrary, it entails an ever increasing sense of responsibility . . . for an ever greater circle of relatedness . . . expressed by the Buddhist term *karunā*—compassion or 'wisdom in action.'" Sponberg's views on compassion are compelling, for they accentuate how we humans need to come to terms with who we are in a hierarchy of compassion. Sponberg also stresses that higher does not mean "better" but rather more responsible.

Love: Toward a paradigm of sensuality and a compassionate new world

> There is a basic goodness in nature. The sun shines. Flowers give fragrance and colour. Fruit gives nourishment. Fire gives warmth. Rain irrigates. There is even simple beauty in winter, death and decay. Nature being red in tooth and claw is a misconception. There is more exuberant beauty in nature than there is cruelty. . . . There is enough in the world for everybody's need, but not enough for anybody's greed.
> —Satish Kumar, "Simplicity for Christmas and Always"

> Compassion—surely that is what the earth seeks most in us.
> —Calvin Luther Martin, *The Way of the Human Being*

If we listen to the spirit of nature, we can elevate our existence on this planet to companion, steward, and "lover." As we learn about other animals and how important they really are to us, we will learn more about ourselves. This knowledge and the intense feelings animals bring forth will help make us nicer to one another and nicer to the planet as a whole. We need to do this now and be proactive, for while I am not a pessimist, I do think that we have limited time. Time is not on our side, mainly because we are so powerful and ubiquitous. The threat of silent springs truly is looming. We must work together to heal a broken and unhealthy world.

On our journeys we will discover that we can indeed love animals more and not love people less. We need to be motivated by love, and not by fear of what it will mean if we come to love animals for who they are. Animals are not less than human. They are who they are and need to be understood in their own worlds.

It all boils down to love. The power of love—the umbrella of love that blankets all beings—must not be underestimated as we try to reconnect with nature and other animals. We need to love Earth and the universe and all of their inhabitants. We need to follow the heat of our hearts and live love. The continued disrespect and abuse of animals, relegating them to being hapless and innocent victims of human greed and arrogance, will make for much loneliness and a severely impoverished universe.

Giving and receiving

In the grand scheme of things, individuals receive what they give. If love is poured out in abundance then it will be returned in abundance. There is no need to fear depleting the potent and self-reinforcing feeling of love that continuously can serve as a powerful stimulant for generating compassion, respect, and more love for all life. Each and every individual plays an essential role and that individual's spirit and love are intertwined with the spirit and love of others. These emergent interrelationships that transcend individuals embodied selves, foster a sense of oneness. These interrelationships can work in harmony to make this a better and more compassionate world for all beings. We must stroll with our kin and not leave them in our tumultuous wake of rampant, self-serving destruction.

It is essential that we do better than our ancestors, and we surely have the resources to do so. Perhaps the biggest question of all is whether enough of us will choose to make the heartfelt commitment to making this a better world, a more compassionate world in which love is plentiful and shared, before it is too late. Will we choose to ignore or respond to a wounded world? Will our complacency get the best of us? Will our fears paralyze us? Will we proceed with grace and humility? I believe we have already embarked on this formidable and necessary pilgrimage. I deeply feel the movement, a pulsating tidal wave, coursing through my body and deep in my heart. My hope and optimism lead me in no other direction.

Giving thanks

Thank you for joining me on my unpredictable and twisting journey. My journey is *our* journey. When we hold hands my hand is your hand. No one can do alone what needs to be done by a community of friends. My transitions and transformations have been many and deep. There have been many beginnings and plenty of false starts. Only a few of my goals have been achieved. This is good, for there is so much exciting, challenging, and enjoyable "work"—proactive passionate activism—to be done. If I make a difference in how humans and animals interact, even a small difference, then my brief residence on Earth, this most amazing planet, will have been well worth the journey.

REFERENCES

Preface

Beck, B. B. 1982. Chimpocentrism: Bias in cognitive ethology. *Journal of Human Evolution* 11:3–17.

Bekoff, M. 2000. *Strolling with our kin: Speaking for and respecting voiceless animals*. New York: Lantern Books.

———, ed. 2000. *The smile of a dolphin: Remarkable accounts of animal behavior*. New York: Random House/Discovery Books. (This book was the subject of cover stories in *U. S. News and World Report*, 30 October 2000, and *USA Today*, 13 December 2000.)

cummings, e. e. 1953. *Six nonlectures*. Cambridge: Harvard University Press.

Raloff, J. 2001. Ill winds. *Science News* 160:218–20.

Rolston, H., III. 1999. *Genesis, genes, and God: Values and their origins in natural and human history*. New York: Cambridge University Press.

Schoen, A. M. 2001. *Kindred spirits: How the remarkable bond between humans and animals can change the way we live*. New York: Broadway Books.

Smith, H. 2001. *Why religion matters*. San Francisco: HarperCollins.

Chapter One

Information on Julia Butterfly Hill and her Circle of Life Foundation can be found at www.circleoflifefoundation.org. The source of Margaret Mead's quotation on activism remains a mystery; see www.mead2001.org/faq=_page.htm#quote.

Allen, C., and M. Bekoff. 1997. *Species of mind: The philosophy and biology of cognitive ethology*. Cambridge: MIT Press.

Beck, A. M., and N. M. Meyers. 1996. Health enhancement and companion animal ownership. *Annual Review of Public Health* 17:247–57.

Bekoff, M., H. L. Hill, and J. B. Mitton. 1975. Behavioral taxonomy in canids by discriminant function analysis. *Science* 190:1223–25.

Boyd, D. K., D. H. Pletscher, and W. G. Brewster. 1993. Evidence of wolves, *Canis lupus*, burying dead wolf pups. *Canadian Field-Naturalist* 107:230–31.

Burkhardt, R. W., Jr. 1981. On the emergence of ethology as a scientific discipline. *Conspectus of History* 1:62–81.

———. 1983. The development of an evolutionary ethology. In *Evolution from molecules to men*, ed. D. S. Bendall, 429–44. New York: Cambridge University Press.

Darwin, C. 1872/1998. *The expression of the emotions in man and animals.* 3d ed., with an introduction, afterword, and commentaries by Paul Ekman. New York: Oxford University Press.

de Waal, F. 2001. *The ape and the sushi master.* New York: Basic Books.

Dewsbury, D. A. 1990. Nikolaas Tinbergen, 1907–1988. *American Psychologist* 45:67–68.

Eisnitz, G. A. 1997. *Slaughterhouse: The shocking story of greed, neglect, and inhumane treatment inside the U.S. meat industry.* Buffalo, N.Y.: Prometheus Books.

Goodall, J. 1990. *Through a window: My thirty years with the chimpanzees of Gombe.* Boston: Houghton Mifflin.

———. 1999. *Reason for hope.* New York: Warner Books.

Haberman, D. 2001. Personal communication, 16 August, at the Third Annual Thomas Berry Symposium, Whidbey Island, Clinton, Washington.

Heinrich, B. 1999. *Mind of the raven: Investigations and adventures with wolf-birds.* New York: Cliff Street Books.

Herzing, D. 2000. A trail of grief. In *The smile of a dolphin: Remarkable accounts of animal emotions*, ed. M. Bekoff, 138–39. New York: Random House/Discovery Books.

Hill, J. B. 2000. *The legacy of Luna.* San Francisco: HarperCollins.

Lorenz, K. Z. 1991. *Here I am—Where are you?* New York: Harcourt Brace Jovanovich.

Mock, D. W., and G. A. Parker. 1997. *The evolution of sibling rivalry.* New York: Oxford University Press.

Moser, A. 2000. The wisdom of nature in integrating science, ethics, and the arts. *Science and Engineering Ethics* 6:365–82.

Pious, S., and H. Herzog. 2001. Reliability of protocol reviews for animal research. *Science* 293:608–9.

Poole, J. 1998. An exploration of a commonality between ourselves and elephants. *Etica & Animali* 9/98:85–110.

Randour, M. L. 2000. *Animal grace: Entering spiritual relationship with our fellow creatures.* Novato, Calif.: New World Library.

Rendell, L., and H. Whitehead. 2001. Culture in whales and dolphins. *Behavioral and Brain Sciences* 24:309–82.

Röell, D. R. 2000. *The world of instinct: Niko Tinbergen and the rise of ethology in the Netherlands, 1920–1950.* Aasen, The Netherlands: Van Gorcum.

Rose, N. 2000. Giving a little latitude. In *The smile of a dolphin: Remarkable accounts of animal emotions*, ed. M. Bekoff, 32. New York: Random House/Discovery Books.

Salem, D. J., and A. N. Rowan, eds. 2001. *The state of the animals 2001.* Washington, D.C.: Humane Society of the United States.

Schusterman, R. J. 2000. Pitching a fit. In *The smile of a dolphin: Remarkable accounts of animal emotions*, ed. M. Bekoff, 106–7. New York: Random House/Discovery Books.

Seattle, Chief. The earth is our mother. www.geocities.com/Athens/6979/prayer.html.

Sewall, L. 1999. *Sight and sensibility: The ecopsychology of perception.* New York: Jeremy P. Tarcher/Putnam.

Tinbergen, N. 1951/1989. *The study of instinct.* New York: Oxford University Press.

———. 1963. On aims and methods of ethology. *Zeitschrift für Tierpsychologie* 20:410–33.

———. 1967. *The herring gull's world.* New York: Anchor Books.

———. 1984. *Curious naturalists.* Amherst: University of Massachusetts Press.

Whiten, A., et al. 1999. Cultures in chimpanzees. *Nature* 399:682–85.

Woolf, M. 1999. Revealed: Secret Slaughter of 9 million "useless" lab animals. *Independent on Sunday,* 15 August, 6.

Chapter Two

Some of the chapter is excerpted or modified from Allen and Bekoff 1997 and from Bekoff 2000, with permission of the respective publishers.

Abram, D. 1996. *The spell of the sensuous: Perception and language in a more-than-human world.* New York: Pantheon.

Adams, E. R., and G. W. Burnett. 1991. Scientific vocabulary divergence among female primatologists working in East Africa. *Social Studies of Science* 21:547–60.

Allen, C., and M. Bekoff. 1997. *Species of mind: The philosophy and biology of cognitive ethology.* Cambridge: MIT Press.

Altmann, S. A. 1967. The structure of primate social communication. In *Social communication among primates,* ed. S. A. Altmann, 325–62. Chicago: University of Chicago Press.

Bekoff, M. 1995. Marking, trapping, and manipulating animals: Some methodological and ethical considerations. In *Wildlife mammals as research models: In the laboratory and field,* ed. K.A.L. Bayne and M. D. Kreger, 31–47. Greenbelt, Md.: Scientists Center for Animal Welfare.

———. 1998. Resisting speciesism and expanding the community of equals. *BioScience* 48:638–41.

———. 2000. Animal emotions: Exploring passionate natures. *BioScience* 50:861–70.

Bekoff, M., and C. Allen. 1997. Cognitive ethology: Slayers, skeptics, and proponents. In *Anthropomorphism, anecdote, and animals: The emperor's new clothes?* ed. R. W. Mitchell, N. Thompson, and L. Miles, 313–34. Albany, N.Y.: SUNY Press.

Bekoff, M., and M. C. Wells. 1996. Social behavior and ecology of coyotes. *Advances in the Study of Behavior* 16:251–338.

Boitani, L., and T. K. Fuller. 2000. *Research techniques in animal ecology: Controversies and consequences.* New York: Columbia University Press.

Burghardt, G. M. 1991. Cognitive ethology and critical anthropomorphism: A snake with two heads and hognose snakes that play dead. In *Cognitive ethology: The minds of other animals—Essays in honor of Donald R. Griffin,* ed. C. A. Ristau, 53–90. Hillsdale, N.J.: Lawrence Erlbaum.

———. Amending Tinbergen: A fifth aim for ethology. In *Anthropomorphism, anecdote, and animals: The emperor's new clothes?* ed. R. W. Mitchell, N. Thompson, and L. Miles, 254–76. Albany, N.Y.: SUNY Press.

de Waal, F. 2001. *The ape and the sushi master.* New York: Basic Books.

Goodall, J. 1999. *Reason for hope*. New York: Warner Books.

Kellert, S. R. 1996. *The value of life: Biological diversity and human society*. Washington: Island Press.

——— . 1997. *Kinship to mastery: Biophilia in human evolution and development*. Washington: Island Press.

——— . 1999. *American perceptions of marine mammals and their management*. Washington, D.C.: Humane Society of the United States.

Kennedy, J. S. 1992. *The new anthropomorphism*. New York: Cambridge University Press.

Lehner, P. N. 1996. *Handbook of ethological methods*. New York: Cambridge University Press.

Macdonald, D. 1987. *Running with the fox*. New York: Facts on File.

McGrew, W. C. 1992. *Chimpanzee material culture: Implications for human evolution*. New York: Cambridge University Press.

Manes, C. 1997. *Other creations: Rediscovering the spirituality of animals*. Garden City, N.Y.: Doubleday.

O'Barry, R. 2000. *Behind the dolphin smile*. Los Angeles: Renaissance Books.

Paul, E. S. 1996. The representation of animals on children's television. *Anthrozoös* 9:169–81.

Phillips, M. T. 1994. Proper names and the social construction of biography: The negative case of laboratory animals. *Qualitative Sociology* 17:119–42.

Poole, J. 2001. Personal communication, 13 June.

Ristau, C. A. 1991. Aspects of the cognitive ethology of an injury-feigning bird, the piping plover. In *Cognitive ethology: The minds of other animals—Essays in honor of Donald R. Griffin*, ed. C. A. Ristau, 91–126. Hillsdale, N.J.: Lawrence Erlbaum.

Rollin, B. E. 1989. *The unheeded cry: Animal consciousness, animal pain, and science*. New York: Oxford University Press (reprinted 1998, Iowa State University Press).

Ryder, R. D. 1998. Speciesism. In *Encyclopedia of animal rights and animal welfare*, ed. M. Bekoff, 320. Westport, Conn.: Greenwood.

Suzuki, D. 2001. Caged animals can go stir crazy. www.canoe.ca/CNEWSScience0110/10-suzuki-can.html.

van den Born, R., et al. 2001. The new biophilia: An exploration of visions of nature in Western countries. *Environmental Conservation* 28: 1–11.

van Krunkelsven, E., J. Dupain, L. van Elsacker, and R. Verheyen. 1999. Habituation of bonobos (*Pan paniscus*): First reactions to the presence of observers and the evolution of response over time. *Folia Primatologica* 70: 365–68.

Varela, F. 2000. Personal communication, 8 December, at a meeting of Science and the Spiritual Quest II, New York City.

Washburn, M. F. 1909. *The animal mind: A text-book of comparative psychology*. London: Macmillan.

White, P.C.L., A. C. Bennett, and E. J. Hayes. 2001. The use of willingness-to-pay approaches in mammal conservation. *Mammal Review* 31: 151–67.

Chapter Three

Alcock, J. 1998. *Animal behavior*. Sunderland, Mass.: Sinauer.

Bagemihl, B. 1999. *Biological exuberance: Animal homosexuality and natural diversity*. New York: St. Martin's Press.

Bekoff, M. 1977. Mammalian dispersal and the ontogeny of individual behavioral phenotypes. *American Naturalist* 111: 715–32.

———. 1995. Vigilance, flock size, and flock geometry: Information gathering by western evening grosbeaks (*Aves fringillidae*). *Ethology* 99: 150–61.

———. 2001. Cunning coyotes: Tireless tricksters, protean predators. In *Model systems in behavioral ecology*, ed. L. Dugatkin, 381–407. Princeton: Princeton University Press.

Bekoff, M., C. Allen, and M. C. Grant. 1999. Feeding decisions by Steller's jays (*Cyanocitta stelleri*): The utility of a logistic regression model for analyses of complex choices. *Ethology* 105: 393–406.

Berger, J., J. E. Swenson, and I.-L. Persson. 2001. Recolonizing carnivores and naive prey: Conservation lessons from Pleistocene extinctions. *Science* 291: 1036–39.

Birkhead, T., and A. Moller, eds. 1998. *Sperm competition and sexual selection*. New York: Academic Press.

Byers, J. A. 1997. *American pronghorn: Social adaptations and the ghost of predators past*. Chicago: University of Chicago Press.

Domb, L. G., and M. Pagel. 2001. Sexual swellings advertise female quality in wild baboon. *Nature* 410: 204–6.

Drea, C. M., J. E. Hawk, and S. E. Glickman. 1996. Aggression decreases as play emerges in infant spotted hyaenas: Preparation for joining the clan. *Animal Behaviour* 51: 1323–36.

Drea, C. M., and K. Wallen. 1999. Low-status monkeys "play dumb" when learning in mixed social groups. *Proceedings of the National Academy of Sciences* 96: 12965–69.

Drickamer, L. C., S. H. Vessey, and D. Meikle. 2001. *Animal behavior: Mechanisms, ecology, and evolution*. Dubuque, Iowa: Wm. C. Brown.

Dugatkin, L. A. 2001. *The imitation factor: Evolution beyond the gene*. New York: Free Press.

———. 2002. *Animal behavior: The interaction of genes, learning, and culture*. New York: W. W. Norton.

Dunbar, R. 1996. *Grooming, gossip, and the evolution of language*. Boston: Faber and Faber.

Etcoff, N. 2000. *Survival of the prettiest: The science of beauty*. New York: Anchor Books.

Hare, B., J. Call, and M. Tomasello. 2001. Do chimpanzees know what conspecifics know? *Animal Behaviour* 61: 139–51.

Harris, S., and P.C.L. White. 1992. Is reduced affiliative rather than increased agonistic behaviour associated with dispersal in red foxes? *Animal Behaviour* 44: 1085–89.

Huffman, M. A. 1997. Current evidence for self-medication in primates: A multidisciplinary perspective. *Yearbook of Physical Anthropology* 40: 171–200.

———. 2001. Self-medicative behavior in the African great apes: An evolutionary perspective into the origins of human traditional medicine. *BioScience* 51: 651–61.

Johnson, V. S. 2000. *Why we feel: The science of emotions*. New York: Perseus Books.

Louie, K., and M. A. Wilson. 2001. Temporally structured replay of awake hippocampal ensemble activity during rapid eye movement sleep. *Neuron* 29: 145–56.

Lozano, G. 1998. Parasitic stress and self-medication in wild animals. *Advances in the Study of Behavior* 27: 291–317.

Pusey, A., J. Williams, and J. Goodall. 1997. The influence of dominance rank on the reproductive success of female chimpanzees. *Science* 277: 828–31.

Ralls, K. 1976. Mammals in which females are larger than males. *Quarterly Review of Biology* 51: 245–76.

Revonsuo, A. 2001. The reinterpretation of dreams: An evolutionary hypothesis of the function of dreaming. *Behavioral and Brain Sciences* 23: 877–901.

Sheldrake, R. 1999. *Dogs that know when their owners are coming home, and other unexplained powers of animals.* London: Hutchinson.

Sheldrake, R., and P. Smart. 2000. Testing a return-anticipating dog. *Anthrozoös* 13: 203–11.

Sober, E. 1984. *The nature of selection.* Cambridge: MIT Press.

Speakman, J., and D. Banks. 1998. The function of flight formations in greylag geese *Anser anser:* Energy savings or orientation? *Ibis* 140: 280–87.

Trivers, R. 2000. William Donald Hamilton, 1936–2000. *Nature* 404: 828.

Wilson, E. O. 1975. *Sociobiology: The new synthesis.* Cambridge: Harvard University Press.

Yudell, M., and R. Desalle. 2000. Sociobiology: Twenty-five years later. *Journal of the History of Biology* 33: 577–84.

Chapter Four

Some of my discussion of consciousness is excerpted and expanded from Allen and Bekoff 1997, with permission of the publisher.

Allen, C. 1998. The discovery of animal consciousness: An optimistic assessment. *Journal of Agricultural and Environmental Ethics* 10: 217–25.

Allen, C., and M. Bekoff. 1997. *Species of mind: The philosophy and biology of cognitive ethology.* Cambridge: MIT Press.

Bateson, P.P.G. 1991. Assessment of pain in animals. *Animal Behaviour* 42: 827–39.

Bekoff, M. 1998. Cognitive ethology: The comparative study of animal minds. In *Blackwell companion to cognitive science,* ed. W. Bechtel and G. Graham, 371–79. Oxford: Blackwell.

Bekoff, M., and C. Allen. 1997. Cognitive ethology: Slayers, skeptics, and proponents. In *Anthropomorphism, anecdote, and animals: The emperor's new clothes?* ed. R. W. Mitchell, N. Thompson, and L. Miles, 313–34. Albany, N.Y.: SUNY Press.

Cheney, D. L., and R. M. Seyfarth. 1990. *How monkeys see the world: Inside the mind of another species.* Chicago: University of Chicago Press.

Dawkins, M. S. 1993. *Through our eyes only? The search for animal consciousness.* San Francisco: W. H. Freeman.

Dennett, D. C. 1983. Intentional systems in cognitive ethology: The "Panglossian paradigm" defended. *Behavioral and Brain Sciences* 6: 343–90.

de Waal, F.B.M. 2001. Pointing primates: Sharing knowledge . . . without language. *Chronicle of Higher Education* (19 January): B7–9.

Dunbar, R.I.M. 1993. The co-evolution of neocortical size, group size, and the evolution of language in humans. *Behavioral and Brain Sciences* 16: 681–735.

Dunbar, R.I.M., and J. Bever. 1998. Neocortex size predicts group size in carnivores and some insectivores. *Ethology* 104: 695–708.

Gallup, G. G., Jr. 1970. Chimpanzees: Self-recognition. *Science* 167: 86–87.

Giurfa, M., et al. 2001. The concepts of "sameness" and "difference" in an insect. *Nature* 410: 930–33.

Griffin, D. R. 1976/1981. *The question of animal awareness: Evolutionary continuity of mental experience.* New York: Rockefeller University Press.

———— . 1992. *Animal minds*. Chicago: University of Chicago Press.

———— . 2001. *Animal minds: Beyond cognition to consciousness*. Chicago: University of Chicago Press.

Hauser, M. 2000. *Wild minds: What animals really think*. New York: Henry Holt.

Heyes, C. 1987. Cognisance of consciousness in the study of animal knowledge. In *Evolutionary epistemology*, ed. W. Callebaut and R. Pinxten, 105–36. Dordrecht: D. Reidel.

Jolly, A. 1966/1988. Lemur social behaviour and primate intelligence. *Science* 153: 501–6. Reprinted in *Machiavellian intelligence*, ed. R. W. Byrne and A. Whiten, 27–33. Oxford: Clarendon Press.

Marler, P. 1996. Social cognition: Are primates smarter than birds? In *Current ornithology volume*, ed. V. Nolan Jr. and E. D. Ketterson, 1–32. New York: Plenum Press.

Matsuzawa, T., ed. 2001. *Primate origins of human cognition and behavior*. New York: Springer.

Mitchell, R. 2002. Kinesthetic visual matching, imitation, and self-recognition. In *The cognitive animal*, ed. M. Bekoff, C. Allen, and G. M. Burghardt, forthcoming. Cambridge: MIT Press.

Nicolakakis, N., and L. Lefebvre. 2000. Forebrain size and innovation rate in European birds: Feeding, nesting and confounding variables. *Behaviour* 137: 1415–29.

Parker, S. T., R. W. Mitchell, and M. L. Boccia, eds. 1994. *Self-awareness in animals and humans: Developmental perspectives*. New York: Cambridge University Press.

Pepperberg, I. 1999. *The Alex studies*. Cambridge: Harvard University Press.

Piggins, D., and C.J.C. Phillips. 1998. Awareness in domesticated animals—Concepts and definitions. *Applied Animal Behaviour Science* 57: 181–200.

Premack, D., and G. Woodruff. 1978. Does the chimpanzee have a theory of mind? *Behavioral and Brain Sciences* 4: 515–26.

Reiss, D., and L. Marino. 2001. Mirror self-recognition in the bottlenose dolphin: A case of cognitive convergence. *Proceedings of the National Academy of Sciences* 98: 5937–42.

Savage-Rumbaugh, E. S. 1997. Why are we afraid of apes with language? In *Origin and evolution of intelligence*, ed. A. B. Scheibel and J. W. Schopf, 43–69. Sudbury, Mass.: Jones and Bartlett.

Sheets-Johnstone, M. 1998. Consciousness: A natural history. *Journal of Consciousness Studies* 5: 260–94.

Shettleworth, S. J. 1998. *Cognition, evolution, and behavior*. New York: Oxford University Press.

Slobodchikoff, C. 2002. Cognition and communication in prairie dogs. In *The cognitive animal*, ed. M. Bekoff, C. Allen, and G. M. Burghardt, forthcoming. Cambridge: MIT Press.

Smuts, B. 1985. *Sex and friendship in baboons*. New York: Aldine.

Tomasello, M., and J. Call. 1997. *Primate cognition*. New York: Oxford University Press.

Yoerg, S. I. 1991. Ecological frames of mind: The role of cognition in behavioral ecology. *Quarterly Review of Biology* 66:287–301.

Zuckerman, S. 1991. Review of Cheney and Seyfarth's *How monkeys see the world: Inside the mind of another species* (1990). *New York Review of Books* (May): 43–49.

Chapter Five

Parts of this chapter are excerpted or modified from M. Bekoff 2000a, with permission of the publisher. The essay in *Der Spiegel* appeared in the 26 March 2001 issue.

Anderson, R. 2000. Seeing red. In *The smile of a dolphin*, 84–87. See M. Bekoff 2000b.

Archer, J. 1999. *The nature of grief: The evolution and psychology of reactions to loss*. New York: Routledge.

Bass, R. 1998a. The wild into the world: An interview with Rick Bass. *International Society for Literature and the Environment* 5:101.

——— . 1998b. *The new wolves: The return of the Mexican wolf to the American southwest*. New York: Lyons Press.

Bekoff, A. 2000. In sickness and in health. In *The smile of a dolphin*, 60–61. See M. Bekoff 2000b.

Bekoff, M. 2000a. Animal emotions: Exploring passionate natures. BioScience 50:861–70.

——— , ed. 2000b. *The smile of a dolphin: Remarkable accounts of animal emotions*. New York: Random House/Discovery Books.

Cabanac, M. 2000. Emotional fever. In *The smile of a dolphin*, 194–97. See M. Bekoff 2000b.

Cheney, D. L., and R. M. Seyfarth. 1990. *How monkeys see the world: Inside the mind of another species*. Chicago: University of Chicago Press.

——— . 1992. Précis of *How monkeys see the world: Inside the mind of another species. Behavioral and Brain Sciences* 15:135–82.

Dagg, A. I. 2000a. Graceful aggression. In *The smile of a dolphin*, 76. See M. Bekoff 2000b.

——— . 2000b. A furious complaint. In *The smile of a dolphin*, 104–5. See M. Bekoff 2000b.

Damasio, A. 1999. *The feeling of what happens: Body and emotion in the making of consciousness*. New York: Harcourt Brace.

Darwin, C. 1859. *On the origin of species by means of natural selection*. London: Murray.

——— . 1871/1936. *The descent of man and selection in relation to sex*. New York: Random House.

——— . 1872/1998. *The expression of the emotions in man and animals*. 3d ed., with an introduction, afterword, and commentaries by Paul Ekman. New York: Oxford University Press.

Dennett, D. C. 1983. Intentional systems in cognitive ethology: The "Panglossian paradigm" defended. *Behavioral and Brain Sciences* 6:343–90.

Frohoff, T. 2000. The dolphin's smile. In *The smile of a dolphin*, 78–79. See M. Bekoff 2000b.

Goodall, J. 1990. *Through a window: My thirty years with the chimpanzees of Gombe*. Boston: Houghton Mifflin.

——— . 2000. Pride goeth before a fall. In *The smile of a dolphin*, 166–67. See M. Bekoff 2000b.

Hauser, M. 2000. A lover's embarrassment? In *The smile of a dolphin*, 200–201. See M. Bekoff 2000b.

Heinrich, B. 1999. *Mind of the raven: Investigations and adventures with wolf-birds*. New York: Cliff Street Books.

——— . 2000. Hopping mad. In *The smile of a dolphin*, 98–99. See M. Bekoff 2000b.

Herzing, D. 1999. Personal communication, 22 November.

Holekamp, K. W., and L. Smale. 2000. A hostile homecoming. In *The smile of a dolphin*, 118–21. See M. Bekoff 2000b.

Lorenz, K. Z. 1991. *Here I am—Where are you?* New York: Harcourt Brace Jovanovich.

McConnery, J. Cited in McRae 2000.

MacLean, P. 1970. *The triune brain in evolution: Role in paleocerebral functions.* New York: Plenum.

McRae, M. 2000. Central Africa's orphaned gorillas: *Will they survive the wild? National Geographic* (February): 86–97.

Masson, J., and S. McCarthy. 1995. *When elephants weep: The emotional lives of animals.* New York: Delacorte Press.

Moss, C. 2000. A passionate devotion. In *The smile of a dolphin,* 135–37. See M. Bekoff 2000b.

Panksepp, J. 2000. The rat will play. In *The smile of a dolphin,* 146–47. See M. Bekoff 2000b.

Pepperberg, I. 2000. Ruffled feathers. In *The smile of a dolphin,* 108. See M. Bekoff 2000b.

Pomiankowski, A., and P. Reguera. 2001. The point of love. *Trends in Ecology and Evolution* 16:533–34.

Poole, J. 1996. *Coming of age with elephants: A memoir.* New York: Hyperion.

———. 1998. An exploration of a commonality between ourselves and elephants. *Etica & Animali* 9/98:85–110.

Rollin, B. 1990. How the animals lost their minds: Animal mentation and scientific ideology. In *Interpretation and explanation in the study of animal behavior,* vol. 1, *Interpretation, intentionality, and communication,* ed. M. Bekoff and D. Jamieson, 375–93. Boulder, Colo.: Westview Press.

Savage-Rumbaugh, S. 2000. Sibling rivalry. In *The smile of a dolphin,* 175. See M. Bekoff 2000b.

Sheets-Johnstone, M. 1998. Consciousness: A natural history. *Journal of Consciousness Studies* 5:260–94.

Siviy, S. 1998. Neurobiological substrates of play behavior: Glimpses into the structure and function of mammalian playfulness. In *Animal play: Evolutionary, comparative, and ecological perspectives,* ed. M. Bekoff and J. A. Byers, 221–42. New York: Cambridge University Press.

Skutch, A. 1996. *The minds of birds.* College Station: Texas A&M University Press.

Smuts, B. 2000. Child of mine. In *The smile of a dolphin,* 150–53. See M. Bekoff 2000b.

Tinbergen, N. 1951/1989. *The study of instinct.* New York: Oxford University Press.

———. 1963. On aims and methods of ethology. *Zeitschrift für Tierpsychologie* 20:410–33.

Tobias, M. 2000. A gentle heart. In *The smile of a dolphin,* 171–73. See M. Bekoff 2000b.

Williams, G. C. 1992. *Natural selection: Domains, levels, and challenges.* New York: Oxford University Press.

Würsig, B. 2000. Leviathan love. In *The smile of a dolphin,* 63–65. See M. Bekoff 2000b.

Chapter Six

Some of this chapter is excerpted and updated from Bekoff 2001, with permission of the publisher. For arguments that we might learn as much or more about the evolution of human social behavior by studying social carnivores, species whose social behavior and organization resemble those of early hominids in a number of ways (divisions of labor, food sharing, care of young, and intersexual and intrasexual dominance hierarchies), see Schaller and Lowther 1969, Tinbergen 1972, and Thompson 1975.

Ackerman, D. 1999. *Deep play.* New York: Random House.

Aldis, O. 1975. *Play fighting*. New York: Academic Press.

Bekoff, M. 1975. The communication of play intention: Are play signals functional? *Semiotica* 15:231–39.

——— . 1977. Social communication in canids: Evidence for the evolution of a stereotyped mammalian display. *Science* 197:1097–99.

——— . 1995. Play signals as punctuation: The structure of social play in canids. *Behaviour* 132:419–29.

——— . 2001. Social play behaviour, cooperation, fairness, trust, and the evolution of morality. *Journal of Consciousness Studies* 8 (2):81–90.

Bekoff, M., and J. A. Byers, eds. 1998. *Animal play: Evolutionary, comparative, and ecological perspectives*. New York: Cambridge University Press.

Bernstein, I. S. 2000. The law of parsimony prevails: Missing premises allow any conclusion. *Journal of Consciousness Studies* 7:31–34.

Brown, S. 1998. Play as an organizing principle: Clinical evidence and personal observations. In *Animal play*, 243–59. See Bekoff and Byers 1998.

Burghardt, G. M. 2002. The *genesis of play*. Cambridge: MIT Press.

Carr, Laurie. 2001. Personal communication, June.

Darwin, C. 1871/1936. *The descent of man and selection in relation to sex*. New York: Random House.

de Waal, F. 1996. *Good-natured: The origins of right and wrong in humans and other animals*. Cambridge: Harvard University Press.

Fagen, R. 1981. *Animal play behavior*. New York: Oxford University Press.

——— . 1993. Primate juveniles and primate play. In *Juvenile primates: Life history, development, and behavior*, ed. M. E. Pereira and L. A. Fairbanks, 183–96. New York: Oxford University Press.

Flack, J. C., and F. de Waal. 2000. Any animal whatever: Darwinian building blocks of morality in monkeys and apes. *Journal of Consciousness Studies* 7:1–29.

Frith, C. D., and U. Frith. 1999. Interacting minds—A biological basis. *Science* 286:1692–95.

Gallese, V. 1998. Mirror neurons, from grasping to language. *Consciousness Bulletin* (Fall): 3–4.

Gallese, V., and A. Goldman. 1998. Mirror neurons and the simulation theory of mind-reading. *Trends in Cognitive Science* 2:493–501.

Gallese, V., P. F. Ferrari, E. Kohler, and L. Fogassi. 2002. The eyes, the hand, and the mind: Behavioral and neurophysiological aspects of social cognition. In *The cognitive animal*, ed. M. Bekoff, C. Allen, and G. Burghardt, forthcoming. Cambridge: MIT Press.

Glausiusz, J. 2001. Our empathic brain. *Discover* (September): 15.

Gruen, L. 2002. The morals of animal minds. In *The cognitive animal*, ed. M. Bekoff, C. Allen, and G. Burghardt, forthcoming. Cambridge: MIT Press.

Hauser, M. 2000. *Wild minds*. New York: Henry Holt.

Mech, L. D. 1970. *The wolf*. Garden City, N.Y.: Doubleday.

Pellis, S. 2002. Keeping in touch: Play fighting and social knowledge. In *The cognitive animal*, ed. M. Bekoff, C. Allen, and G. Burghardt, forthcoming. Cambridge: MIT Press.

Peterson, G. R. 2000. God, genes, and cognizing agents. *Zygon* 35:469–80.

Power, T. G. 2000. *Play and exploration in children and animals*. Hillsdale, N.J.: Lawrence Erlbaum.

Ridley, M. 1996. *The origins of virtue: Human instincts and the evolution of cooperation.* New York: Viking.

Schaller, G. B., and G. R. Lowther. 1969. The relevance of carnivore behavior to the study of early hominids. *Southwestern Journal of Anthropology* 25:307–41.

Spinka, M., R. C. Newberry, and M. Bekoff. 2000. Mammalian play: Training for the unexpected. *Quarterly Review of Biology* 76:141–68.

Thompson, P. R. 1975. A cross-species analysis of carnivore, primate, and hominid behavior. *Journal of Human Evolution* 4:113–24.

Tinbergen, N. 1972. Introduction to H. Kruuk, *The spotted hyena.* Chicago: University of Chicago Press.

Watson, D. M., and D. B. Croft. 1996. Age-related differences in playfighting strategies of captive male red-necked wallabies (*Macropus rufogriseus banksianus*). *Ethology* 102:333–46.

Wechlin, S., J. H. Masserman, and W. Terris Jr. 1964. Shock to a conspecific as an aversive stimulus. *Psychonomic Science* 1:17–18.

Wickler, W. 1972. *The biology of the ten commandments.* New York: McGraw-Hill.

Chapter Seven

Much of this chapter is excerpted (and in some places updated) from Bekoff 2000. Current views that question the value of animal models in medical research can be found in Greek and Greek 2000. Information on the United Kingdom's ban on great apes in research can be found at paragraphs 10–12 of the Supplementary Note to the Home Secretary's response to the Animal Procedures Committee—Interim Report on the Review of the Operation of the Animals (Scientific Procedures) Act 1986, 6 November 1997. The Sunday *Times* (London, 8 April 2001, 70) reported that Oxford University and Imperial College (London) were avoiding the government ban by having their research conducted at the Biomedical Primate Research Centre (BPRC) in Rijswijk, The Netherlands. The article in the *Washington Post* titled "They Die Piece by Piece" can be found at www.washingtonpost.com/wp-dyn/articles/A60798-2001Apr9.html. More information on nonanimal alternatives in education can be found at www.hsus.org/programs/research/annotate.html.

Achor, A. B. 1996. *Animal rights: A beginner's guide.* Yellow Springs, Ohio: WriteWare.

Ammann, K. 2001. Bushmeat hunting and the great apes. In *Great apes and humans: The ethics of coexistence,* ed. B. Beck et al., 71–85. Washington: Smithsonian Institution Press.

Balcombe, J. 2000. *The use of animals in education: Problems, alternatives, and recommendations.* Washington: Humane Society of the United States.

Balls, M., A.-M. van Zeller, and M. E. Halder, eds. 2000. *Progress in the reduction, refinement, and replacement of animal experimentation.* The Netherlands: Elsevier.

Bateson, P.P.G. 1997. The behavioural and physiological effects of culling red deer. London: Report to the Council of the National Trust.

Beck, A., G. F. Melson, P. L. da Costa, and T. Liu. 2001. The educational benefits of a ten-week home-based feeding program for children. *Anthrozoös* 14:19–28.

Beck, B. B. 1996. Reintroduction of captive-bred animals. In *The well-being of animals in zoo and aquarium sponsored research,* ed. G. M. Burghardt et al., 61–65. Greenbelt, Md.: Scientists Center for Animal Welfare.

Bekoff, M. 2000. *Strolling with our kin: Speaking for and respecting voiceless animals.* New York: Lantern Books.

———, ed. 1998. *Encyclopedia of animal rights and animal welfare.* Foreword by Jane Goodall. Westport, Conn.: Greenwood.

Bekoff, M., and D. Jamieson. 1996. Ethics and the study of carnivores. In *Carnivore behavior, ecology, and evolution,* ed. J. L. Gittleman, 16–45. Ithaca: Cornell University Press.

Bentham, J. 1996. *An introduction to the principles of morals and legislation,* chapter 17, note 1. Oxford: Clarendon Press.

Bostock, S. St. C. 1993. *Zoos and animal rights.* London: Routledge.

Bowen-Jones, E., and S. Pendry. 1999. The threat to primates and other mammals from the bushmeat trade in Africa, and how this threat could be diminished. *Oryx* 33:233–46.

Campbell, T. C., and C. J. Chen. 1994. Diet and chronic degenerative diseases: Perspectives from China. *American Journal of Clinical Nutrition* 59:1153–61.

Croke, V. 1997. *The modern ark: The story of zoos, past, present, and future.* New York: Scribner.

Davis, K. 1996. *Poisoned chickens, poisoned eggs: An inside look at the modern poultry industry.* Summertown, Tenn.: Book Publishing Company.

Duda, M. D., S. J. Bissell, and K. C. Young. 1996. Factors related to hunting and fishing participation in the United States. *Transactions of the 61st American Wildlife and Natural Resources Conference,* 324–37.

Dunlap, T. R. 1988. *Saving America's wildlife: Ecology and the American mind.* Princeton: Princeton University Press.

Eisnitz, G. A. 1997. *Slaughterhouse: The shocking story of greed, neglect, and inhumane treatment inside the U.S. meat industry.* Buffalo, N.Y.: Prometheus Books.

Enck, J. W., D. J. Decker, and T. L. Brown. 2000. Status of hunter recruitment and retention in the United States. *Wildlife Society Bulletin* 4:817–24.

Fouts, R., with S. Mills. 1997. *Next of kin: What chimpanzees have taught me about who we are.* New York: William Morrow.

Fox, M. A. 1999. *Deep vegetarianism.* Philadelphia: Temple University Press.

Francione, G. L. 2000. *Introduction to animal rights: Your child or the dog?* Philadelphia: Temple University Press.

Goodall, J. 1987. A plea for chimpanzees. *American Scientist* 75:574–77.

———. 1990. *Through a window: My thirty years with the chimpanzees of Gombe.* Boston: Houghton Mifflin.

Greek, R., and J. Greek. 2000. *Sacred cows and golden geese.* New York: Continuum.

Green, A. 1999. *Animal underworld: Inside America's market for rare and exotic species.* New York: PublicAffairs.

Hancocks, D. 2001. *A different nature: The paradoxical world of zoos and their uncertain future.* Berkeley: University of California Press.

Holsman, R. H. 2000. Goodwill hunting: Exploring the role of hunters as ecosystem stewards. *Wildlife Society Bulletin* 28:816.

Jolly, A. 1991. Conscious chimpanzees? A review of recent literature. In *Cognitive ethology: The minds of other animals—Essays in honor of Donald R. Griffin,* ed. C. A. Ristau, 231–52. Hillsdale, N.J.: Lawrence Erlbaum.

Mason, G. J., J. Cooper, and C. Clarebrough. 2001. Frustrations of fur-farmed mink. *Nature* 410:35–36.

Mighetto, L. 1990. *Wild animals and American environmental ethics.* Tucson: University of Arizona Press.

Nelson, R. 1991. *The island within.* New York: Vintage Books.

Regan, T. 1983. *The case for animal rights.* Berkeley: University of California Press.

Rifkin, J. 1992. *Beyond beef: The rise and fall of the cattle culture.* New York: Dutton.

Robinson, M. 2001. Adapt or perish? Zoos must choose (review of Hannocks). *Science* 292:1304–5.

Russell, W.M.S., and R. L. Burch. 1959/1992. *The principles of humane experimental technique.* Wheathampstead, England: UFAW.

Samsel, R. W., G. A. Schmidt, J. B. Hall, L.D.H. Wood, S. G. Shroff, and P. T. Schumaker. 1994. Cardiovascular physiology teaching: Computer simulations vs. animal demonstrations. *Advances in Physiology Education* 11:536–46.

Sapontzis, S. 1995. We should not allow dissection of animals. *Journal of Agricultural and Environmental Ethics* 8:181–89.

Seligman, M.E.P., S. F. Maier, and J. H. Geer. 1968. Alleviation of learned helplessness in the dog. *Journal of Abnormal Psychology* 73:256–62.

Shepherdson, D. J., J. D. Mellen, and M. Hutchins, eds. 1998. *Second nature: Environmental enrichment for captive animals.* Washington: Smithsonian Institution Press.

Singer, P. A. 1990. *Animal liberation.* New York: Random House.

Wilcove, D. S., D. Rothstein, J. Dubow, A. Phillips, and E. Losos. 1998. Quantifying threats to imperiled species in the United States. *BioScience* 48:607–15.

Wilkie, D. S. 2001. Bushmeat trade in the Congo Basin. In *Great apes and humans: The ethics of coexistence,* 86–109. Washington: Smithsonian Institution Press.

Wise, S. M. 2000. *Rattling the cage: Towards legal rights for animals.* Cambridge, Mass.: Perseus Books.

Woodroffe, R., J. Ginsberg, and D. Macdonald. 1997. *The African wild dog.* Gland, Switzerland: International Union for Conservation of Nature and Natural Resources.

Zinko, U., N. Jukes, and C. Gericke. 1997. *From guinea pig to computer mouse: Alternative methods for a humane education.* The Netherlands: EuroNiche.

Chapter Eight

Some material in this chapter has been excerpted and updated from Bekoff 2000, with permission of the publisher. The sections on why study animals and moving on are adapted from Bekoff and Jamieson 1996.

Baldi, A., and P. Betáry. 1999. Microclimate and vegetation edge effects in a reedbed in Hungary. *Biodiversity and Conservation* 8:1697–706.

Bekoff, M. 2000. Field studies and animal models: The possibility of misleading inferences. In *Progress in the reduction, refinement, and replacement of animal experimentation,* ed. M. Balls, A.-M. van Zeller, and M. E. Halder, 1553–59. The Netherlands: Elsevier.

Bekoff, M., and D. Jamieson. 1991. Reflective ethology, applied philosophy, and the moral status of animals. *Perspectives in Ethology* 9:1–47.

——— . 1996. Ethics and the study of carnivores: Doing science while respecting animals. In *Carnivore behavior, ecology, and evolution,* ed. J. L. Gittleman, 16–45. Ithaca: Cornell University Press.

Byers, J. A. 1998. *American pronghorn: Social adaptations and ghosts of predators past.* Chicago: University of Chicago Press.

Eisemann, C. H., et al. 1984. Do insects feel pain? *Experientia* 40:164–67.

Ferrer, M., and F. Hiraldo. 1992. Man-induced sex-biased mortality in the Spanish imperial eagle. *Biological Conservation* 60:57–60.

Fox, C. H., and C. M. Papouchis. 2002. *The cull of the wild: The politics of trapping in the United States.* Sacramento: Animal Protection Institute.

Hauser, M. D. 1993. Do vervet monkey infants cry wolf? *Animal Behaviour* 45:1242–44.

Huntingford, F. A. 1984. Some ethical issues raised by studies of predation and aggression. *Animal Behaviour* 32:210–15.

Janss, G.F.E., A. Lazo, and M. Ferrer. 1999. Use of raptor models to reduce avian collisions with powerlines. *Journal of Raptor Research* 33:154–59.

Kenney, S. P., and R. L. Knight. 1992. Flight distances of black-billed magpies in different regimes of human density and persecution. *Condor* 94:545–47.

Laurenson, M. K., and T. M. Caro. 1994. Monitoring the effects of non-trivial handling in free-living cheetahs. *Animal Behaviour* 47:547–57.

Chapter Nine

Some of this chapter has been excerpted and updated from Bekoff 2000, with permission of the publisher.

Berry, T. 1999. *The great work.* New York: Bell Tower.

Bekoff, M. 1999. Lynx and academic freedom. *Boulder Camera* (22 July): 7A (www.bouldernews.com/opinion/columnists/bekmarc.html).

——— . 2000. Redecorating nature: Reflections on science, holism, humility, community, reconciliation, spirit, compassion, and love. *Human Ecology Review* 7:59–67.

——— . 2001. Human-carnivore interactions: Adopting proactive strategies for complex problems. In *Carnivore conservation,* ed. J. L. Gittleman, S. M. Funk, D. W. Macdonald, and R. K. Wayne, 179–95. London: Cambridge University Press.

Berger, J., A. Hoylman, and W. Weber. 2001. Perturbation of vast ecosystems in the absence of adequate science: Alaska's arctic refuge. *Conservation Biology* 15:539–41.

Berkes, F. 1999. *Sacred knowledge: Traditional ecological knowledge and resource management.* Philadelphia: Taylor and Frances.

BioScience. 2001. Scientific objectivity, value systems, and policymaking. June, special issue.

Caro, T. M., ed. 1998. *Behavioral ecology and conservation biology.* New York: Oxford University Press.

Clark, T. W., M. Stevenson, K. Ziegelmayer, and M. Rutherford, eds. 2001. *Species and ecosystem conservation: An interdisciplinary approach.* New Haven: Yale University Press.

Courchamp, F., and D. W. Macdonald. 2001. Crucial importance of pack size in the African wild dog *Lycaon pictus. Animal Conservation* 4:169–74.

Cowlishaw, G., and R. Dunbar. 2000. *Primate conservation biology.* Chicago: University of Chicago Press.

Crabtree, R. L., and J. W. Sheldon. 1999. Coyotes and canid coexistence in Yellowstone. In *Carnivores in ecosystems,* ed. T. W. Clark, A. P. Curlee, S. C. Minta, and P. M. Kareiva, 127–63. New Haven: Yale University Press.

Crooks, K. R., and M. E. Soulé. 1999. Mesopredator release and avifaunal extinctions in a fragmented system. *Nature* 400:563–66.

East, M. L., and H. Hofer. 1996. Wild dogs in the Serengeti ecosystem: What really happened? *Trends in Ecology and Evolution* 11:509.

Ehrlich, P. R. 1997. *A world of wounds: Ecologists and the human dilemma.* Oldendorg/ Luhe, Germany: Ecology Institute.

Ehrlich, P. R. 2000. *Human natures: Genes, cultures, and the human prospect.* Washington: Island Press.

Estes, J. A. 1998. Concerns about rehabilitation of oiled wildlife. *Conservation Biology* 12:1156–57.

Ginsberg, J. R., K. A. Alexander, S. Creel, P. W. Kat, J. W. McNutt, and G. L. Mills. 1995. Handling and survivorship of African wild dog (*Lycaon pictus*) in five ecosystems. *Conservation Biology* 9:665–74.

Glendinning, C. 1999. *Off the map.* Boston: Shambhala.

Gosling, L. M., and W. J. Sutherland, eds. 2000. *Behaviour and conservation.* New York: Cambridge University Press.

Gruchow, P. 1988. *The necessity of open places.* New York: St. Martin's Press.

Kay, C. E., B. Patton, and C. A. White. 2000. Historical wildlife observations in the Canadian Rockies: Implications for ecological integrity. *Canadian Field-Naturalist* 114:561–83.

Kirkwood, J. 1992. Wild animal welfare. In *Animal welfare and the environment,* ed. D. Ryder and P. Singer, 139-54. London: Duckworth.

Kloor, K. 1999. Lynx and biologists try to recover after disastrous start. *Science* 285:320–21.

Leopold, A. 1949. *A Sand County almanac.* New York: Oxford University Press.

McKibben, B. 1995. An explosion of green. *Atlantic Monthly* 275 (April): 61–83.

McKinney, M. L. 2001. Role of human population size in raising bird and mammal threat among nations. *Animal Conservation* 4:45–57.

Nader, L., ed. 1996. *Naked science: Anthropological inquiry into boundaries, power, and knowledge.* New York: Routledge.

Oates, J. F., et al. 2000. Extinction of a West African red colobus monkey. *Conservation Biology* 14:1526–32.

Posey, D. A., ed. 1999. *Cultural and spiritual values of biodiversity.* Nairobi, Kenya: United Nations Environment Programme.

Quinn, D. 1993. *Ishmael.* New York: Bantam.

Sutherland, W. J. 1998. The importance of behavioural studies in conservation biology. *Animal Behaviour* 56:801–9.

Suzuki, D., and H. Dressel. 1999. *From naked ape to super-species.* New York: Stoddart.

Tear, T., M. Scott, P. Hayward, and B. Griffith. 1993. Status and prospects for success of the Endangered Species Act: A look at recovery plans. *Science* 262:976–77.

——— . 1995. Recovery plans and the Endangered Species Act: Are criticisms supported by data? *Conservation Biology* 9:182–95.

Terborgh, J., J. A. Estes, P. Paquet, K. Ralls, D. Boyd-Heger, B. Miller, and R. F. Noss. 1999. The role of top carnivores in regulating terrestrial ecosystems. In *Continental conservation: Scientific foundations of regional reserve networks,* ed. M. E. Soulé and J. Terborgh, 39–64. Washington: Island Press.

Webb, N. R. 1999. Ecology and ethics. *Trends in Ecology and Evolution* 14:259–60.

White, D., Jr., K. C. Kendall, and H. D. Picton. 1999. Potential energetic effects of mountain climbers on foraging grizzly bears. *Wildlife Society Bulletin* 27:146–51.

White, N. 1999. Lynx, free speech tangle at CU. *Boulder Camera* (17 July): 1B, 4B (www.buffzone.com/buffzone/news/17cflap.html).

Woodroffe, R., and J. R. Ginsberg. 1997. Past and future causes of wild dogs' population decline. In *The African wild dog,* ed. R. Woodroffe, J. Ginsberg, and D. Macdonald, 58–74. Gland, Switzerland: International Union for Conservation of Nature and Natural Resources.

Chapter Ten

Anderson, E. N. 1996. *Ecologies of the heart: Emotion, belief, and the environment.* New York: Oxford University Press.

Birke, L. 1994. *Feminism, animals, and science: The naming of the shrew.* Buckingham, England: Open University Press.

———. 1997. Science and animals—or why Cyril won't win the Nobel Prize. *Animal Issues* 1:45–55.

Clayton, P. 2001. Review of Mary Midgley, *Science and poetry. Nature* 409:979–80.

Corning, P. A. 1983. *The synergism hypothesis: A theory of progressive evolution.* New York: McGraw-Hill.

Ehrenfeld, D. 1981. *The arrogance of humanism.* New York: Oxford University Press.

Fisher, G. 2001. Sustainable human development: Connecting the scientific and moral dimensions. Unpublished manuscript.

Fox, W. 1990. *Toward a transpersonal ecology.* Boston: Shambhala.

Goodall, J., and M. Bekoff. 2002. *Ten trusts.* San Francisco: HarperCollins.

Griffin, D. R. 2000. *Religion and scientific naturalism.* Albany, N.Y.: SUNY Press.

Gunther, P.A.Y. 2000. Leopold's land ethics, Texas, and the big thicket: An obligation to the land. *Texas Journal of Science* 52 (supplement): 23–32.

Jolly, A. 1999. *Lucy's legacy: Sex and intelligence in evolution.* Cambridge: Harvard University Press.

Kumar, S. 2000. Simplicity for Christmas and always. *Resurgence* (November/December): 3.

Lack, D. 1957. *Evolutionary theory and Christian belief.* London: Methuen.

Lorimer, D. 1999. Introduction: From experiment to experience. In *The spirit of science: From experiment to experience,* ed. D. Lorimer, 17–29. New York: Continuum.

McElroy, S. C. 1995. *Animals as teachers and healers.* Troutdale, Ore.: NewSage Press.

Martin, C. L. 1999. *The way of the human being.* New Haven: Yale University Press.

Midgley, M. 1992. *Science as salvation: A modern myth and its meaning.* New York: Routledge.

———. 2001. *Science and poetry.* New York: Routledge.

Newberg, A., D. D'Aquili, and V. Rause. 2001. *Why God won't go away.* New York: Ballantine.

Regenstein, L. G. 1991. *Replenish the earth: A history of organized religion's treatment of animals and nature, including the Bible's message of conservation and kindness toward animals.* New York: Crossroad.

Sagan, C. 1997. *Billions and billions.* New York: Random House.

Sponberg, A. 1997. Green Buddhism and the hierarchy of compassion. In *Buddhism and ecology,* ed. M. E. Tucker and D. R. Williams, 351–76. Cambridge: Harvard University Press.

Thomas, E. M. 1996. *Certain poor shepherds: A Christmas tale.* New York: Simon and Schuster.

Tucker, M. E., and J. A. Grim, eds. 1994. *Worldviews and ecology: Religion, philosophy, and the environment.* Maryknoll, N.Y.: Orbis Books.

Williams, T. T. 2001. *Red: Passion and patience in the desert.* New York: Pantheon Books.

Wilson, E. O. 1998. *Consilience: The unity of knowledge.* New York: Alfred A. Knopf.

INDEX

Note: Page numbers in italics refer to illustrations.